Coupled Modes

IN

Plasmas, Elastic Media,

AND

Parametric Amplifiers

•

A Numerical Method

Modern Analytic *and* Computational Methods *in* Science *and* Mathematics

A GROUP OF MONOGRAPHS AND ADVANCED TEXTBOOKS

Richard Bellman, EDITOR
University of Southern California

Already Published:

1. R. E. Bellman, Harriet H. Kagiwada, R. E. Kalaba, and Marcia C. Prestrud, Invariant Imbedding and Radiative Transfer in Slabs of Finite Thickness, 1963
2. R. E. Bellman, Harriet H. Kagiwada, and Marcia C. Prestrud, Invariant Imbedding and Time-Dependent Transport Processes, 1964
3. R. E. Bellman and R. E. Kalaba, Quasilinearization and Nonlinear Boundary-Value Problems, 1965
4. R. E. Bellman, R. E. Kalaba, and Jo Ann Lockett, Numerical Inversion of the Laplace Transform: Applications to Biology, Economics, Engineering, and Physics, 1966
5. S. G. Mikhlin and K. L. Smolitskiy, Approximate Methods for Solution of Differential and Integral Equations, 1967
6. R. N. Adams and E. D. Denman, Wave Propagation and Turbulent Media, 1966
7. R. L. Stratonovich, Conditional Markov Processes and Their Application to the Theory of Optimal Control, 1968
8. A. G. Ivakhnenko and V. G. Lapa, Cybernetics and Forecasting Techniques, 1967
9. G. A. Chebotarev, Analytical and Numerical Methods of Celestial Mechanics, 1967
10. S. F. Feshchenko, N. I. Shkil', and L. D. Nikolenko, Asymptotic Methods in the Theory of Linear Differential Equations, 1967
11. A. G. Butkovskiy, Distributed Control Systems, 1969
12. R. E. Larson, State Increment Dynamic Programming, 1968
13. J. Kowalik and M. R. Osborne, Methods for Unconstrained Optimization Problems, 1968
14. S. J. Yakowitz, Mathematics of Adaptive Control Processes, 1969
15. S. K. Srinivasan, Stochastic Theory and Cascade Processes, 1969
16. D. U. von Rosenberg, Methods for the Numerical Solution of Partial Differential Equations, 1969
17. R. B. Banerji, Theory of Problem Solving: An Approach to Artificial Intelligence, 1969
18. R. Lattès and J.-L. Lions, The Method of Quasi-Reversibility: Applications to Partial Differential Equations. Translated from the French edition and edited by Richard Bellman, 1969
19. D. G. B. Edelen, Nonlocal Variations and Local Invariance of Fields, 1969
20. J. R. Radbill and G. A. McCue, Quasilinearization and Nonlinear Problems in Fluid and Orbital Mechanics, 1970
24. D. H. Jacobson and D. Q. Mayne, Differential Dynamic Programming, 1970
27. E. D. Denman, Coupled Modes in Plasmas, Elastic Media, and Parametric Amplifiers, 1970

In Preparation:

21. W. Squire, Integration for Engineers and Scientists, 1970
22. T. Parthasarathy and T. E. S. Raghavan, Some Topics in Two-Person Games
23. T. Hacker, Flight Stability and Control, 1970
25. H. Mine and S. Osaki, Markovian Decision Processes, 1970
26. W. Sierpiński, 250 Problems in Elementary Number Theory
28. F. A. Northover, Applied Diffraction Theory
29. G. A. Phillipson, Identification of Distributed Systems
30. D. H. Moore, Heaviside Operational Calculus: An Elementary Foundation
31. S. M. Roberts and J. S. Shipman, Two-Point Boundary Value Problems: Shooting Methods

Coupled Modes
IN
Plasmas, Elastic Media,
AND
Parametric Amplifiers

A Numerical Method

•

Eugene D. Denman

Department of Electrical Engineering
UNIVERSITY OF HOUSTON
Houston, Texas

American Elsevier
Publishing Company, Inc.
NEW YORK · 1970

AMERICAN ELSEVIER PUBLISHING COMPANY, INC.
52 Vanderbilt Avenue, New York, N.Y. 10017

ELSEVIER PUBLISHING COMPANY, LTD.
Barking, Essex, England

ELSEVIER PUBLISHING COMPANY
335 Jan Van Galenstraat, P.O. Box 211
Amsterdam, The Netherlands

Standard Book Number 444-00074-7

Library of Congress Card Number 70-104978

Copyright © 1970 by American Elsevier Publishing Company, Inc.

All rights reserved.
No part of this publication may be reproduced,
stored in a retrieval system, or transmitted
in any form or by any means, electronic,
mechanical, photocopying, recording,
or otherwise, without the prior
written permission of the publisher,
American Elsevier Publishing Company, Inc.,
52 Vanderbilt Avenue, New York, N.Y. 10017.

Printed in the United States of America

THIS BOOK IS DEDICATED

TO THE MEMORY OF

James Spencer Davis

●

Contents

CHAPTER 1

Introduction

CHAPTER 2

Scattering Matrices, Normal Modes, and State Variables

1.	Scattering Matrices.	10
2.	Transmission Matrices	14
3.	Normal Mode Theory.	16
4.	Normal Mode Equations and the Transmission Line	18
5.	State Variables and State Space	20
	References.	22

CHAPTER 3

Recursive Equations

1.	The Single-Mode Problem.	23
2.	The Multimode Problem	30
3.	The Recursive Equations and the Continuous Medium	34
4.	Medium Thickness Increasing in the Direction of Negative x.	36
5.	Change in the Direction of Flow.	37
6.	The Matrix Product Formulation	39
7.	Recursive Relations for Forcing Functions.	41
8.	Summary	44
	References.	44

CHAPTER 4

The Riccati, Recursive, and Other Differential Equations

1.	Homogeneous Second-Order Differential Equations and the Coupled First-Order Differential Equations.	47
2.	The Coupled First-Order Equations and the Riccati Equation	48
3.	The Recursive Equations and the Riccati Equation	51

4. Differential Equations of the Matrix Product 53
5. Differential Equations for the Transmission Matrix when $u(x)$ is the "Flow" to the Left. 55
6. Initial Value and Linear Two-Point Boundary-Value Problems 57
7. Differential Equations for the Source Functions 59
8. The Inhomogeneous Differential Equation and the Matrix Product Formulation. 60
9. Integration in the Negative x Direction 62
10. The Incremental Coefficients for the T Matrix 65
11. Incremental Coefficients for the Source Functions 66
12. The Incremental Coefficients for the Fundamental Matrix: $u(x)$ in Direction of $x+$, $x > 0$ 67
13. Recursive Equations for the Integral of the Inhomogeneous Differential Equation 70
14. Differential Equations with Periodic Solutions: Initial Value Problems. 72
15. Properties of the Fundamental Matrix 73
16. The Numerical Solution of One First-Order Differential Equation 74
17. Summary 75
References. 76

CHAPTER 5

Solution of Differential Equations: Initial Value Problems

1. First-Order Differential Equations 78
2. A Network Transient Problem 81
3. The Airy Differential Equation 83
4. Integration of Functions 86
5. Nonlinear Differential Equations with Initial Values . . . 89
6. Inhomogeneous Differential Equations 91
7. A Matrix Differential Equation 93
8. Numerical Calculations of Eigenvalues 100
9. Summary 100
References. 101

CHAPTER 6

Two-Point Boundary-Value Problems

1. Linear Two-Point Boundary-Value Problem 102
2. Mixed Linear Two-Point Boundary-Value Problems . . . 103
3. Quasilinearization of Nonlinear Differential Equations . . . 104

4. Solutions of Nonlinear Two-Point Boundary-Value Problems . 106
5. Example of a Linear Two-Point Boundary-Value Problem . . 106
6. Nonlinear Two-Point Boundary-Value Problems 107
7. Summary 110
Reference 111

CHAPTER 7

Singular Values and Integration in the Complex Plane

1. Unbounded Incremental Coefficients. 113
2. Singular Values of the Matrix Coefficients 114
3. Large Transmission Matrix Coefficients and Computational Errors 116
4. Integration in the Complex Plane 117
5. The Complex Fresnel Integral 120
6. Summary 122
References 122

CHAPTER 8

The Recursive Procedure Compared with Other Numerical Methods

1. Runge-Kutta Method 124
2. An Example Problem for Comparing the Runge-Kutta and the Invariant Imbedding Methods 126
3. The Adam-Moulton Method 129
4. Numerical Comparison of Several Equations 130
5. Summary 132
References 133

CHAPTER 9

Transmission Matrices for Circuits and Discrete Media

1. Time-Invariant Components and the Scattering Matrix . . . 134
2. Multimodes in Linear Circuits 139
3. Transmission Matrix for Interacting Modes in a Series-Nonlinear Element 141
4. Transmission Matrix for Interacting Modes in a Shunt-Nonlinear Element 145
5. The Transmission Matrix for a Layered Isotropic Medium: Single Mode at Normal Incidence 146

6. The Continuous Medium and Electromagnetic Waves at Normal
   Incidence 148
7. Oblique Incidence Case 149
8. Summary 149
References 150

CHAPTER 10

Plane Elastic Waves in a Lossless Inhomogeneous Medium

1. Propagation in an Unbounded Medium 152
2. Interface Coefficients 158
3. Conservation of Energy 162
4. Numerical Results 163
5. Summary 168
References 169

CHAPTER 11

Magnon-Phonon Coupling in Ferrites

1. Magnon-Phonon Equations of Motion in Ferrites 170
2. The Coupled First-Order Differential Equations 172
3. Values of the Constants 174
4. Computational Results 175
5. Summary 179
References 181

CHAPTER 12

Polarization of an Electromagnetic Wave in an Inhomogeneous Ionized Medium

1. The Inhomogeneous Ionized Medium 183
2. The Coupled Mode Equations 186
3. The Booker Quartic, Polarization, and the Direction of Propagation 191
4. Determination of the Boundary Conditions 193
5. Some Numerical Results 195
6. Summary 200
References 200

CHAPTER 13

Electromagnetic-Electroacoustic Wave Coupling in a Compressible Plasma

1.	The Ionized Medium and the Linearized Plasma Equations	202
2.	The Method of Analysis	205
3.	The Boundary Conditions and the Interface Coefficients	206
4.	The Ionized Layer Model	210
5.	Numerical Results	210
6.	Summary	216
References		216

CHAPTER 14

Parametric Amplifiers

1.	Simple Amplifier	218
2.	Negative Resistance, Up Conversion, and Down Conversion Gain Equations	222
3.	The Gain Equations	224
4.	Summary	227

APPENDICES

A.	Fourth-Order Incremental Coefficients	228
B.	Sample Computer Program	231
Author Index		233
Subject Index		235

List of Tables

5.1.	Solution of $u'(x) = v(x)$ with initial value $u(0) = 1$	80
5.2.	Transient problem—single loop.	82
5.3.	Airy functions	86
5.4.	Error function	88
5.5.	Nonlinear equation: $u'(x) = -2xu^2(x)$	90
5.6.	Inhomogeneous coupled equations.	93
5.7.	Network elements and roots	97
5.8.	Matrix equation	99
5.9.	Tabulated critical lengths	100
6.1.	Solution of equation (6.6)	107
6.2.	Tabulated values of nonlinear two-point boundary-value problem	109
7.1.	Transport equation	115
7.2.	Computational error $u(x)$	117
7.3.	Complex transport problem.	119
7.4.	Complex Fresnel integral	121
8.1.	Pendulum problem	128
8.2.	Differential equation with constant coefficients	131
8.3.	$(1 + x^2)^{-1}$ solution of equation (8.14).	132
10.1.	Elastic constants for water, sandstone, and limestone	163
11.1.	Magnon and phonon amplitudes with $m(0) \neq 0$, $r(0) \neq 0$	177
11.2.	Magnon and phonon amplitudes with $m(0) = 0$, $r(0) \neq 0$	179
12.1.	Constants for sample problems.	186
13.1.	Imaginary components of electromagnetic and electrostatic wave numbers.	211
13.2.	Reflection coefficients for lossy Epstein layer: Maximum electron density = 1.49172×10^8 electrons/cc.	211
13.3.	Transmission coefficients for lossy Epstein layer: Maximum density = 1.49172×10^8 electrons/cc	212
13.4.	Reflection coefficients for lossy Epstein layer: Maximum electron density = 10^{10} electrons/cc	213
13.5.	Transmission coefficients for lossy Epstein layer: Maximum electron density = 10^{10} electrons/cc	213

List of Illustrations

Figure 2.1.	One-port network	11
Figure 2.2.	Two-port network	13
Figure 2.3.	Transmission matrix of two-port network	14
Figure 2.4.	Simple oscillator	16
Figure 2.5.	Two coupled modes	18
Figure 2.6.	Section of transmission line	19
Figure 2.7.	Block diagram for state equations	21
Figure 3.1.	Transmission line of discrete sections	24
Figure 3.2.	A simple interface	26
Figure 3.3.	Flow graph for simple interface	27
Figure 3.4.	Single-slab model	27
Figure 3.5.	Signal flow graph for single slab	28
Figure 3.6.	Coupled mode model	31
Figure 3.7.	Flow graph for single-slab coupled model	32
Figure 3.8.	Continuous line	35
Figure 3.9.	Signal flow graph for negative x	37
Figure 3.10.	Signal flow graph for opposite flow	38
Figure 3.11.	Signal flow graph with sources	42
Figure 3.12.	Cascade signal flow graph with sources	43
Figure 4.1.	Boundary condition model	49
Figure 4.2.	Incremental flow for reverse flow	56
Figure 4.3.	Integration in the direction of negative x	63
Figure 5.1.	Simple network	81
Figure 5.2.	Two-loop network	94
Figure 6.1.	Two solutions of nonlinear equations	110
Figure 7.1.	Deformed contour	118
Figure 8.1.	Simple pendulum with constant torque	126
Figure 9.1.	Two-port network	135
Figure 9.2.	Shunt impedance	136
Figure 9.3.	Series impedance	137
Figure 9.4.	Shunt-series two-port network	138
Figure 9.5.	Two-port network with N waves	139
Figure 9.6.	Series nonlinear capacitor	142

Figure 9.7.	Shunt nonlinear capacitor	145
Figure 9.8.	Abrupt impedance change	146
Figure 10.1.	Forces acting on a parallelepiped	155
Figure 10.2.	Interaction of P and SV waves at a solid-solid interface	159
Figure 10.3.	Two-interface elastic medium	163
Figure 10.4.	Transmission coefficients for varying angles of incidence	164
Figure 10.5.	Reflection coefficients for varying angles of incidence	164
Figure 10.6.	Three-layer solid	165
Figure 10.7.	Transmission coefficients for varying sandstone slab thickness	166
Figure 10.8.	Reflection coefficients for varying sandstone slab thickness	167
Figure 10.9.	Transmission coefficients for varying limestone slab thickness	167
Figure 10.10.	Reflection coefficients for varying limestone slab thickness	168
Figure 11.1.	Ferrite medium	171
Figure 11.2.	Phonon-magnon coupling	176
Figure 11.3.	Phonon-magnon coupling	178
Figure 11.4.	Magnon amplitudes vs. distance	180
Figure 11.5.	Phonon amplitudes vs. distance	180
Figure 12.1.	Square of the refractive index vs. $X(z)$	184
Figure 12.2.	Square of the refractive index vs. $X(z)$	184
Figure 12.3.	Epstein profile	185
Figure 12.4.	Reflection process for ionized layer	187
Figure 12.5.	R_s vs. angle of incidence	196
Figure 12.6.	R_s vs. angle of incidence	196
Figure 12.7.	R_s vs. angle of incidence	197
Figure 12.8.	Relative coupled power vs. angle of incidence	198
Figure 12.9.	Relative coupled power vs. angle of incidence	199
Figure 12.10.	Percent of extraordinary wave power coupled to ordinary wave vs. angle of incidence	199
Figure 13.1.	Stratified ionized medium	205
Figure 13.2.	Wave branching at an interface in ionized medium	207
Figure 13.3.	Transmissivity of electromagnetic wave vs. angle of incidence	214
Figure 13.4.	Transmissivity of electromagnetic wave vs. angle of incidence	215
Figure 13.5.	Absorption factor for lossy Epstein profile	215
Figure 14.1.	Simple parametric amplifier.	218

Preface

This book describes a numerical method for solving coupled mode problems which are frequently encountered in mathematical physics and engineering. The method is based on the recursive equations of invariant imbedding. Coupled mode problems, described by linear and nonlinear coupled first-order differential equations or in a discrete sense, can be solved by the procedure. Although the "wave" analogy is used in the method and a selected group of wave propagation are discussed, the method is not limited to wave problems. It can be utilized in solving differential equations, definite integrals, optimal control problems, and other engineering problems.

The method should be of interest to engineers, as well as to applied mathematicians working on engineering-type problems. The "wave" analogy which is used makes it possible to associate physical significance with the differential equations. The numerical scheme is not difficult to develop and the method can be easily understood by an engineer with a limited training in mathematics.

A detailed development of the numerical method is given in the book. The connection between the scattering matrix and the numerical method has been given. The numerical method is discussed and numerous differential equations are solved to illustrate the procedure and to verify the accuracy of the scheme. Several wave propagation problems are solved and described; they are of the type in which interaction between various waves occurs within the medium.

The coupled mode problems, which can be solved by the method, are not limited to initial value problems; two-point boundary-value problems can also be analyzed. This makes the method useful in solving many problems that cannot be solved by other numerical methods.

EUGENE D. DENMAN

Nashville, Tennessee
Spring 1970

CHAPTER I

Introduction

There are many mathematical problems in engineering and physics which can be described by the differential equations

$$\frac{du(x)}{dx} = B(x)u(x) + A(x)v(x) + e(x), \tag{1.1}$$

$$-\frac{du(x)}{dx} = D(x)u(x) + C(x)v(x) + f(x). \tag{1.2}$$

Equations such as these are considered in the book and a *numerical method* for solving this coupled set is described. The problem, which is described by the set of equations, can be either an initial or a two-point boundary-value one.

A "wave" analogy is postulated and the mathematical functions $u(x)$ and $v(x)$ are treated as waves. The transmission matrix for a medium characterized by the coefficients $A(x)$, $B(x)$, $C(x)$, and $D(x)$ will be defined and calculated. A recursive method is used to compute the transmission matrix. The "waves," $u(x)$ and $v(x)$, are constructed from the calculated transmission matrix.

To give the reader a preview of the method, let us consider (1.1) and (1.2) as describing the waves $u(x)$ and $v(x)$ in an inhomogeneous medium. Let $u(x)$ be a wave propagating in the positive x direction, and $v(x)$ be a wave moving in the negative x direction. We assume that an inhomogeneous medium of thickness x is imbedded in a free-space region of fine extent. Since $u(x)$ is a wave moving in the direction of positive x, $u(0)$ is the incident wave on the medium at $x = 0$. The wave incident on the medium at x is $v(x)$. The wave $u(x)$ consists of that part of $u(0)$ transmitted through the medium, a portion of $v(x)$ that is reflected, and a contribution from internal sources within the medium. In a similar manner, $v(0)$ is made of a transmitted component of $v(x)$, a reflected wave from $u(0)$, and a wave due to sources in the medium. Since the inhomogeneous medium is imbedded in a region of infinite extent, waves which emerge from the inhomogeneous space do not undergo further reflections in the free space.

This wave picture makes it possible to define "wave" equations for $u(x)$ and $v(x)$. The wave which emerges from the medium at x is

$$u(x) = \tau(0, x)u(0) + R(0, x)v(x) + g(0, x),$$

whereas the wave leaving the medium at $x = 0$ satisfies the equation

$$v(0) = \mathscr{R}(0, x)u(0) + T(0, x)v(x) + h(0, x).$$

The functions $\tau(0, x)$, $R(0, x)$, $\mathscr{R}(0, x)$, and $T(0, x)$ are elements of the transmission matrix, and the source contributions to the waves are expressed by the functions $g(0, x)$ and $h(0, x)$.

The two preceding equations are essential to the development of the numerical method outlined in this chapter. Mathematical details on development of the numerical method are given in the chapters which follow.

The transmission matrix and the source functions can be calculated in a recursive manner. To illustrate the calculation of these functions, consider a medium of zero thickness imbedded in free space. The incident waves on each side of the medium is completely transmitted, and no wave can be excited by internal sources. The initial conditions for the transmission matrix and source functions are

$$T(0, 0) = \tau(0, 0) = I,$$
$$R(0, 0) = \mathscr{R}(0, 0) = g(0, 0) = h(0, 0) = 0.$$

If the inhomogeneous medium of any thickness is extended by an infinitesimal thickness Δx, the transmission matrix and source contributions obey the six equations below.

$$R(0, x + \Delta x) = R(x, x + \Delta x) + \tau(x, x + \Delta x)$$
$$\times [I - R(0, x)\mathscr{R}(x, x + \Delta x)]^{-1} \cdot R(0, x)T(x, x + \Delta x),$$
$$\tau(0, x + \Delta x) = \tau(x, x + \Delta x)[I - R(0, x)\mathscr{R}(x, x + \Delta x)]^{-1}\tau(0, x),$$
$$T(0, x + \Delta x) = T(0, x)[I - \mathscr{R}(x, x + \Delta x)R(0, x)]^{-1}T(x, x + \Delta x),$$
$$\mathscr{R}(0, x + \Delta x) = \mathscr{R}(0, x) + T(0, x)[I - \mathscr{R}(x, x + \Delta x)R(0, x)]^{-1}$$
$$\cdot \mathscr{R}(x, x + \Delta x)\tau(0, x),$$
$$g(0, x + \Delta x) = g(x, x + \Delta x) + \tau(x, x + \Delta x)$$
$$\times [I - R(0, x)\mathscr{R}(x, x + \Delta x)]^{-1}$$
$$\cdot [g(0, x) + R(0, x)h(x, x + \Delta x)],$$
$$h(0, x + \Delta x) = h(0, x) + T(0, x)[I - \mathscr{R}(x, x + \Delta x)R(0, x)]^{-1}$$
$$\cdot [h(x, x + \Delta x) + \mathscr{R}(x, x + \Delta x)g(0, x)].$$

These six recursive equations can be used for analyzing a medium of any thickness. If the medium is of zero thickness, the initial conditions and the

I INTRODUCTION

recursive equations can be used to extend the medium to a thickness Δx. This medium is then increased in thickness by an additional Δx and the process continued until the desired thickness is obtained.

To use these six equations in an algorithm the transmission matrix and source functions for an infinitesimal thickness Δx must be defined. To find the first-order approximations for these functions, which are called the incremental coefficients, consider the Taylor series expansions for $u(x + \Delta x)$ and $v(x + \Delta x)$.

$$u(x + \Delta x) = u(x) + \frac{du(x)}{dx} \Delta x + \cdots,$$

$$v(x + \Delta x) = v(x) + \frac{dv(x)}{dx} \Delta x + \cdots.$$

Substituting for $du(x)/dx$ and $dv(x)/dx$ from (1.1) and (1.2), and considering first-order terms,

$$u(x + \Delta x) = [I + B(x) \Delta x]u(x) + A(x) \Delta x v(x + \Delta x) + e(x) \Delta x,$$
$$v(x) = D(x) \Delta x u(x) + [I + C(x) \Delta x]v(x + \Delta x) + f(x) \Delta x.$$

These equations are valid for a thickness Δx and, by comparing these equations with those for a thickness Δx, the first-order incremental coefficients are

$$\tau(x, x + \Delta x) = I + B(x) \Delta x,$$
$$T(x, x + \Delta x) = I + C(x) \Delta x,$$
$$R(x, x + \Delta x) = A(x) \Delta x,$$
$$\mathscr{R}(x, x + \Delta x) = D(x) \Delta x,$$
$$g(x, x + \Delta x) = e(x) \Delta x,$$
$$h(x, x + \Delta x) = f(x) \Delta x.$$

The recursive equations can now be utilized in finding the six functions necessary for calculating $u(x)$ and $v(x)$ everywhere. Assume that $u(0)$ and $v(0)$ are specified, then $u(x)$ can be calculated directly from the two "wave" equations. Although the discussion above has been greatly simplified, the recursive equations are quite useful.

Higher-order incremental coefficients are developed in the book, and the fourth-order coefficients are given in Appendix A. A sample program is given in Appendix B; it indicates the simplicity of using the method.

It may appear to the reader that the "wave" analogy is an artificial operation since the solutions can be found directly from the Taylor series development. The analogy is a very useful means of obtaining a physical picture of the coupling process. Furthermore, the transmission matrix provides a method of attacking nonlinear two point boundary-value problems, as well as solving linear and nonlinear initial value problems.

This book describes the numerical method based on the wave picture and the recursive equations. The method is related to the principle of invariance introduced by Bellman and Kalaba [1] and by Redheffer [2]. Bellman has used the term "invariant imbedding" to make a distinction between the method and other analytical methods of mathematical physics. The author has used the method of invariant imbedding in a previous book [3] as well as in numerous papers. Since the mathematical foundation of invariant imbedding has been developed in the above-mentioned [1–2], as well as by Wing [4] and Preisendorfer [5], only the application to problems is presented here. The method described, since it is related to invariant imbedding, is frequently referred to as an invariant imbedding method.

The transmission matrix, which is utilized in the numerical scheme, is basically the scattering matrix. The reader may be aware of the obvious relationship between the transmission matrix and the scattering matrix; should he not be, some introductory material is included. Many of the properties of the scattering matrix can be applied directly to problems in transmission lines and circuits. A brief discussion of normal modes is also included because of the obvious connection to the work here. Readers acquainted with scattering matrices and normal mode theory may choose to omit the introductory material on these subjects in Chapter 2.

The recursive equations for single and multimode problems are derived in Chapter 3. Signal flow graphs are used to derive the recursive equations for the transmission matrix, as well as for the source terms. Since the recursive equations are of major importance in all of the work, considerable effort is spent on the derivation of these equations. Once the proper foundation has been developed for the transmission matrix, the matrix product formulation is described. The matrix product describes the fundamental matrix [6], or the state transition matrix [7] which has found extensive use in state space analysis. The behavior of a physical system can be defined by state equations, i.e.,

$$\frac{d\mathbf{X}(t)}{dt} = \mathbf{A}\mathbf{X}(t) + \mathbf{B}\mathbf{Z}(t), \tag{1.3}$$

$$\mathbf{Y}(t) = \mathbf{C}\mathbf{X}(t) + \mathbf{D}\mathbf{Z}(t). \tag{1.4}$$

Equation (1.3) is, in general, a matrix equation of order n, and this equation can easily be modified such that two equations in coupled mode form are considered.

I INTRODUCTION

Chapter 4 is devoted to the derivation of the differential equations for the transmission matrix elements. Similarly, the four differential equations for the matrix product formulation are given. The transmission matrix elements obey nonlinear differential equations. It is shown that the $n \times n$ transmission matrix can be partitioned into four matrices of order $n/2 \times n/2$: two transmission matrix coefficients, and two reflection matrix coefficients. One of the reflection coefficients obeys the Riccati equation. This equation is not solved by applying a known numerical scheme; it is shown that the recursive equation for this reflection coefficient is a discrete version of the Riccati equation. It is also shown that all of the differential equations, those of the fundamental and transmission matrices, can be solved by using recursive equations.

The Riccati differential equation is a rather interesting equation since it appears in so many areas of mathematical physics. Because the differential equation does not have an analytical solution, a numerical scheme must be used to solve it. The Riccati equation always appears in determining the reflection coefficient of a wave on a transmission line having a spatial variation of impedance. The reflection coefficient is related to the impedance of the line; see P. Wang [8]. A rather interesting report showing the connection between the Riccati equation and network theory has been written by Kaplan and Stock [9].

The Riccati equation also appears in control theory. If a control system is initiated in a certain state while under the influence of a forcing function, and must reach a certain terminal state in a prescribed time, the Riccati equation must be solved. The solution to the Riccati equation describes how the system must perform between the initial and final states.

Descriptions of control problems of this type are given by Sage [10], Friedland [11], Porter [12], and others. A computing method for linear optimum control problems based on invariant imbedding is also given by Tai et al. [13]. The reader interested in these problems may find the numerical scheme presented here useful in optimum control problems.

The solution of linear differential equations with initial conditions is discussed in Chapter 5. The recursive method is illustrated by solving homogeneous and inhomogeneous differential equations. A simple first-order differential equation is first discussed to introduce the reader to the method. This particular problem has been selected to test the accuracy of the recursive equations and is a rather trivial equation. More complicated equations are solved in this chapter. All of the problems selected have known solutions or have been tabulated in published form elsewhere. Although the problems cannot be classified as coupled mode problems, such as occur in mathematical physics, the problems are useful in illustrating the method.

The Airy differential equation is discussed in Chapter 5. This equation is of interest for several reasons. The equation not only provides a means of checking the numerical accuracy of the method but also gives a method by which fractional-order Bessel functions can be generated. The results of the calculations were compared to the published values of Smirnov [14]. The Airy differential equation is a second-order equation, i.e.,

$$L(u) = u''(x) + [\lambda^2 x r_2(x) + q(x)]u(x) = 0. \tag{1.5}$$

The equations for $u'(x)$ and $u''(x)$ were used to generate the coupled mode equations.

The numerical scheme is not limited to differential equations. Definite integrals of the type

$$u(x) = \int_0^x e(\alpha)\, d\alpha \tag{1.6}$$

can be solved numerically. If $A(x) = B(x) = 0$, then (1.6) is the integral equivalent of (1.1). Although the integral is not in a coupled mode form, it is not difficult to show that the recursive equation for $g(0, x)$ holds and the solution is given by

$$u(x) = u(0) + g(0, x), \tag{1.7}$$

where $\tau(0, x) = 1$, and $R(0, x) = \mathcal{R}(0, x) = 0$. The recursive equation for $g(0, x + \Delta x)$ is then given by

$$g(0, x + \Delta x) = g(x, x + \Delta x) + g(0, x). \tag{1.8}$$

The integral of the error function,

$$H(x) = \frac{2}{\sqrt{\pi}} \int_0^x e^{-\alpha}\, d\alpha, \tag{1.9}$$

is investigated in Chapter 5. The differential equation formulation was used rather than the forcing function form. Ten-decimal accuracy was obtained for this particular integral.

The numerical method is not limited to linear differential equations of the types given in (1.1) and (1.2). Nonlinear differential equations with initial conditions can be solved by the method. The nonlinear coupled mode equations can be expressed in the forms

$$\frac{du(x)}{dx} = B(u, v, x)u(x) + A(u, v, x)v(x) + e(x), \tag{1.10a}$$

$$-\frac{dv(x)}{dx} = D(u, v, x)u(x) + C(u, v, x)v(x) + f(x). \tag{1.10b}$$

I INTRODUCTION

Since the solutions for $u(x)$ and $v(x)$ can be found for each infinitesimal thickness Δx, the coefficients $B(u, v, x)$, $A(u, v, x)$, $D(u, v, x)$, and $C(u, v, x)$ can be calculated everywhere.

A simple first-order nonlinear differential equation of the form

$$\frac{dy(x)}{dx} = -2xy^2(x) \tag{1.11}$$

was considered, and the results are given in Chapter 5.

The recursive equations are directly applicable to a system of n equations, where $A(x)$, $B(x)$, $C(x)$, and $D(x)$ are matrices. A two-loop electrical network is described and analyzed in Chapter 5. Four simultaneous equations are solved; this is the highest-order system of equations discussed in the book. Higher-order systems have been solved by the method with excellent results.

Many systems which are encountered in engineering, such as optimal control problems, are described as two-point boundary-value problems. Assume that the initial value $u(0)$ is given, and suppose that the value of $v(x)$ is given at $x = a$. It is not difficult to show that the linear two-point boundary-value problem can be solved by making two passes through the medium. The six recursive equations are solved for the entire medium, the missing value of $v(0)$ is found, and the solutions are constructed on the second pass. Nonlinear differential equations must be solved by an iterative means. Quasilinearization is first applied to the nonlinear differential equations. The recursive equations are then used, along with an initial guess for $v(0)$, to calculate the six functions and the approximate values of $u(x)$ and $v(x)$. A better approximation is found when the entire medium has been analyzed; this value is then used for calculating a second approximate solution. A new value of $v(0)$ is found and the process continued until the solution is found. Problems of this type are described in Chapter 6.

Chapter 7 discusses some of the difficulties which may arise in using the method. If any of the coefficients $A(x)$, $B(x)$, $C(x)$, and $C(x)$ are singular at any point, the incremental coefficients cannot be defined. Methods of avoiding these singular points must be included in the numerical scheme. A limited discussion of this troublesome feature is given, as well as a means of avoiding singularities in the transmission matrix. Since integration in the complex plane is valid, the latter problem may be eliminated by making a contour into the complex plane. Integration in the complex plane has been used with success, and several examples are discussed in Chapter 7.

An attempt has been made in Chapter 8 to compare the invariant imbedding method with the Runge-Kutta and Adams-Moulton methods. The invariant imbedding method has an accuracy comparable to that of these methods when the incremental coefficients are accurately described. Since a Taylor series development of the incremental coefficients is used, the series must converge.

Chapter 9 discusses a few methods of deriving the transmission matrix for simple circuits and discrete medium. The methods presented are by no means the only ways to obtain the reflection, coupling, and transmission coefficients. Some of the problems considered in the last part of the book illustrate other means of obtaining the incremental coefficients.

The remainder of the book is devoted to some of the problems considered in theses and dissertations submitted by graduate students at Vanderbilt University. Propagation of elastic waves in an elastic medium, magnon-phonon interaction in a ferrite, electromagnetic wave propagation in an ionized media, and the parametric amplifier are described. All of these problems have been analyzed using the numerical method given in this book.

The problems considered in the final chapters are simplified models. Surface effects for a medium of finite dimensions have not been included; all of the problems consider the medium to be infinite along two of the axes.

Much of the material contained in this book has been presented to graduate students at Vanderbilt University over the past years. Constructive criticisms made by the students have been extremely helpful and their effort in all of the work is appreciated. The author expresses his thanks to many of these students; in particular, to Dr. Roy Adams, Dr. James Kelley, and Edward McKay. The last four chapters have been taken from the work of Dr. Hugh Davies, Winsor Letton, III, Raymond Dodd, and William Bradley. The book is dedicated to Mr. James S. Davis, a student who will be long remembered by the author.

The author is grateful to the National Science Foundation, and to Vanderbilt University for support during much of this work. The preliminary work was started under NSF Grant No. GP 3117 and continued under Grant No. GK 1413. It would have been impossible for the research to have been completed without their support.

The manuscript for the book was partially written at the University of Southern California, when the author was on leave from Vanderbilt University. Encouragement given by Dr. Richard E. Bellman has been helpful and many suggestions of Dr. Allan Schumitzky have been extremely useful.

REFERENCES

1. R. Bellman and R. Kalaba, On the Principle of Invariant Imbedding and propagation through an Inhomogeneous media, *Proc. Natl. Acad. Sci. U.S.*, 42 (1956) 629.
2. R. Redheffer, Difference Equations and Functional Equations in Transmission-Line Theory, *Modern Mathematics for the Engineer*, Second Series, McGraw-Hill Book Co., New York, 1962, 282.
3. R. N. Adams and E. D. Denman, *Wave Propagation and Turbulent Media*, American Elsevier, New York, 1967.
4. G. M. Wing, *An Introduction to Transport Theory*, John Wiley and Sons, New York, 1962.
5. R. Preisendorfer, Invariant Imbedding Relations for the Principle of Invariance, *Proc. Natl. Acad. Sci. U.S.*, 44 (1958), 320.
6. R. Struble, *Nonlinear Differential Equations*, McGraw-Hill Book Co., New York, 1962.
7. L. Zadeh and C. Desoer, *Linear System Theory: The State Space Approach*, McGraw-Hill Book Co., New York, 1963.
8. P. Wang, Impedance of a Linear System with Anisotropic Coefficients, *J. Franklin Inst.*, 284 (1967), 161.
9. L. J. Kaplan and D. J. R. Stock, Interpretation of Riccati Equation for Some Circuits, Tech. Rept. 400-38, Dept. of E. E., New York University, New York, 1961.
10. A. Sage, *Optimum Systems Control*, Prentice-Hall Inc., Englewood Cliffs, N.J., 1968.
11. B. Friedland, On Solutions of the Riccati Equation in Optimization Problems, IEEE *Trans. on Automatic Controls* (1967), 303.
12. W. Porter, On the Matrix Riccati Equation, IEEE *Trans. on Automatic Controls*, AC-12 (1967), 746.
13. Tai Ju Wei, Li Pow So, and Wang Yu Yin, A Computing Method for the Linear Time Optimal Control System, *Acta Automatica Servica*, 2 (1966), 123. (Translation available from U.S. Dept. of Commerce, Scientific and Technical Information, FTD-TT-65-1921.)
14. A. D. Smirnov, *Table of Airy Functions and Special Confluent Hypergeometric Functions*, Pergamon Press, New York, 1960.

CHAPTER 2

Scattering Matrices, Normal Modes, and State Variables

The scattering matrix method of analyzing scattering problems in networks is well known to many electrical engineers. Carlin and Giordano [1] have written a comprehensive text on the mathematical theory of scattering matrices as applied to networks. The transmission matrix used in the invariant imbedding theory that follows does not differ greatly from the scattering matrix. The only difference is in the definition of the wave. Because of the duality which exists between the scattering and transmission matrices, it is felt that a review of some of the basic properties of the scattering matrix would be of interest and help to introduce the transmission matrix. Because some of the work presented later makes use of the scattering matrix, the reader should have a limited knowledge of the scattering matrix.

As will be shown, the "wave" of scattering matrix theory is a linear combination of two variables or "waves." Since this is the mathematical foundation of normal mode theory, the two topics are discussed in this chapter. Although the invariant imbedding technique does not depend on normalization of the "waves" or generation of a modified wave, invariant imbedding and normal mode theory have much in common. An attempt is made here to present the material to reveal the similarities; much remains to be done to determine the mathematical connection between the two methods.

The normal mode theory leads, in a somewhat natural way, to the study of the Riccati equation. Since this is the basic equation of the work in this book, some of the invariant imbedding techniques carry over directly to normal mode theory. The reader should consult the excellent book by Louisell [2] for more details on normal mode theory.

I. SCATTERING MATRICES

The conventional variables in a linear network, as analyzed by the impedance or admittance formalism, are voltage and current. These variables are

2 SCATTERING MATRICES, NORMAL MODES, STATE VARIABLES

the first ones that are introduced to students in the study of electrical circuits; these variables are used in the two-loop network problem analyzed in a later chapter.

In analyzing a network by the scattering matrix method, a linear combination of voltage and current is assumed. The new set of variables are the incident and reflected waves. Since a linear combination of voltage and current is used to construct the waves, normalization of the voltage and current variables can be made to give the correct dimensions to the waves. To illustrate the scattering matrix concept, let us consider a one-port network as

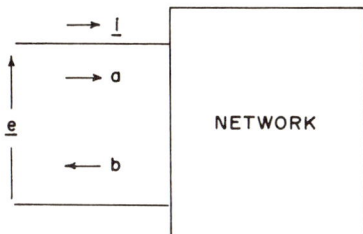

Figure 2.1. One-port network.

shown in Figure 2.1. The one-port network is constructed of lumped-passive circuit elements. The bar under the voltage $\underline{e}$ and the current $\underline{i}$ signifies that these variables are nonnormalized variables. The distinction between normalized and nonnormalized variables is not always made in this book since the normalization is arbitrary.

Assume a linear combination of voltage and current such that the incident wave a can be taken as

$$a = \alpha_{11}\underline{e} + \alpha_{12}\underline{i} \tag{2.1a}$$

and the reflected wave b as

$$b = \alpha_{21}\underline{e} + \alpha_{22}\underline{i}. \tag{2.1b}$$

The constants α_{ij} are usually chosen to make a and b convenient for the physical problem being considered. Generally, these constants are chosen to make the power transfer equations as simple as possible. The usual choice of the constants is

$$\alpha_{11} = \alpha_{21} = \frac{1}{2\sqrt{r_0}}, \quad \alpha_{12} = -\alpha_{22} = \frac{\sqrt{r_0}}{2}, \tag{2.2}$$

where $\sqrt{r_0}$ is a normalizing constant. If r_0 has the dimensions of ohms, then a and b have the dimensions of (volts-amperes)$^{1/2}$.

Using the constants defined in (2.2), (2.1a) and (2.1b) become

$$a = \frac{1}{2\sqrt{r_0}} [\underline{e} + i\underline{r_0}], \tag{2.3a}$$

$$b = \frac{1}{2\sqrt{r_0}} [\underline{e} - i\underline{r_0}]. \tag{2.3b}$$

If the normalized voltage and current are now defined as

$$\underline{e} = e\sqrt{r_0}, \qquad \underline{i} = i\sqrt{r_0},$$

then

$$a = \tfrac{1}{2}[e + i], \tag{2.4a}$$

$$b = \tfrac{1}{2}[e - i], \tag{2.4b}$$

No restrictions are placed on the normalizing constants; thus they can be selected at the discretion of the individual. The normalizing constants given above are the most common ones.

The ratio of reflected to incident voltage for the one-port network is defined as the reflection coefficient, generally denoted by S, where

$$S = \frac{b}{a}. \tag{2.5}$$

The transmission coefficient T can be defined by considering the remainder of the wave. That portion of the wave not reflected must be transmitted to the load. If S is the reflection coefficient, then

$$T = 1 - S = \frac{a - b}{a}. \tag{2.6}$$

Utilizing (2.4) and (2.6), we find that S and T can be defined as functions of the voltage and current:

$$S = \frac{e - i}{e + i} = \frac{Z - 1}{Z + 1}, \tag{2.7a}$$

$$T = \frac{2i}{e + i} = \frac{2}{Z + 1}, \tag{2.7b}$$

where Z is the normalized impedance, defined as $\underline{Z} = Zr_0$.

2 SCATTERING MATRICES, NORMAL MODES, STATE VARIABLES

The one-port network has a single reflection coefficient, whereas the two-port network must be defined in terms of a 2 × 2 scattering matrix. Each of the two ports possesses a reflection coefficient, and there must be a transmission coefficient for each port. Let us consider the two-port network given in Figure 2.2.

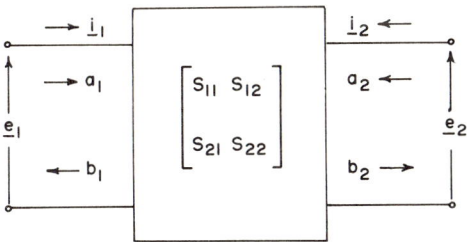

Figure 2.2. Two-port network.

If a_i is the incident wave at the ith port and b_i the reflected wave, then a_i and b_i can be defined as a linear combination of e_i and i_i:

$$a_1 = \frac{1}{2\sqrt{r_{01}}} [e_1 + r_{01} i_1], \tag{2.8a}$$

$$b_1 = \frac{1}{2\sqrt{r_{01}}} [e_1 - r_{01} i_1], \tag{2.8b}$$

$$a_2 = \frac{1}{2\sqrt{r_{02}}} [e_2 + r_{02} i_2], \tag{2.8c}$$

$$b_2 = \frac{1}{2\sqrt{r_{02}}} [e_2 - r_{02} i_2], \tag{2.8d}$$

where r_{01} and r_{02} are the normalizing constants for port 1 and 2, respectively. The two-port network has the scattering matrix

$$\begin{bmatrix} b_1 \\ b_2 \end{bmatrix} = \begin{bmatrix} S_{11} & S_{12} \\ S_{21} & S_{22} \end{bmatrix} \begin{bmatrix} a_1 \\ a_2 \end{bmatrix} \tag{2.9}$$

or, in matrix notation,

$$b = [S]a, \tag{2.10}$$

where a and b are column vectors.

The scattering matrix is valid for a multiport network. If the network has n ports, then S will be a $n \times n$ matrix, a and b will be column vectors of order n.

2. TRANSMISSION MATRICES

The scattering matrix above considers the "wave" b_i as a sum of the reflected "wave" from one of the ports and the transmitted "wave" from the other port. A simple transformation can be made to define the outgoing waves in terms of the transmitted waves. Let us consider a change in the network of Figure 2.2 as shown in Figure 2.3. Let T be a 2×2 matrix such that

$$\begin{bmatrix} b_1 \\ b_2 \end{bmatrix} = \begin{bmatrix} \mathscr{R} & T \\ \tau & R \end{bmatrix} \begin{bmatrix} a_1 \\ a_2 \end{bmatrix} \qquad (2.11)$$

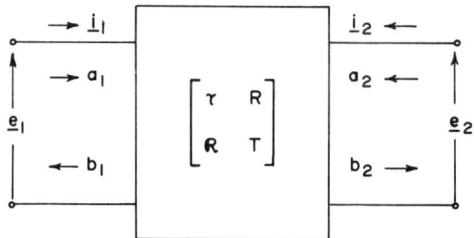

Figure 2.3. Transmission matrix of two-port network.

but, since b_1 of Figure 2.3 is equal to b_1 of Figure 2.2, and b_2 of Figure 2.3 is equal to b_2 of Figure 2.2, we find that

$$\mathbf{T} = \begin{bmatrix} \tau & R \\ \mathscr{R} & T \end{bmatrix} = \begin{bmatrix} S_{21} & S_{22} \\ S_{11} & S_{12} \end{bmatrix}. \qquad (2.12)$$

It is a simple matter to show that the matrix transformation required to obtain (2.12) from [S] is

$$[\mathbf{T}] = \begin{bmatrix} 0 & I \\ I & 0 \end{bmatrix} [\mathbf{S}] = [\Omega][\mathbf{S}]. \qquad (2.13)$$

Ω is hereafter called the transformation matrix.

The transformation between [T] and [S] is useful in the study of parametric amplifiers. The scattering matrix can be found for the circuit elements of which the amplifier is comprised, and the transmission matrix can be immediately found. Equation (2.13) is valid for a single-mode problem. When a multimode problem is considered, the [S] matrix must be rearranged to obtain the correct vector form. The column vectors $\mathbf{a}_1$, $\mathbf{a}_2$, $\mathbf{b}_1$, and b_2 must correspond to the input vector on the left side, the input vector on the right side, the reflected vector wave on the left side, and the reflected vector wave on the right side, respectively.

Let S be the modified scattering matrix such that it is written for the vectors discussed above. Equation (2.10) is still valid, since a_1, a_2, b_1 and b_2 are now vectors. The transformation equation (2.13) is correct where I is now the identity matrix. This transformation is discussed in more detail in Chapter 9. The reader should use care in setting up the S matrix before making the transformation. The input and output vectors should be correctly defined.

The **T** matrix given in (2.12) is henceforth called the transmission matrix. A second form of scattering matrix, frequently encountered in the literature, may confuse the reader if he is not aware of its significance. Frequent use of the $[ABCD]$ parameters is made in network theory when two-port networks are to be cascaded. It is convenient to use the $ABCD$ parameters since the cascading operation can be expressed through a straight matrix multiplication. Unfortunately, the scattering matrix for a two-port network does not permit the cascading operation of several two-port networks to be described by the matrix multiplication. To circumvent this problem, the transfer scattering matrix (see Carlin and Giordano, [1]), has been defined. The transfer scattering matrix is nothing more than the fundamental or state transition matrix. If $[\mathbf{S}(j)]$ is the scattering matrix for the jth two-port network, then a matrix $\mathbf{P}(j)$† can be defined such that

$$\begin{bmatrix} b_1(1) \\ a_1(1) \end{bmatrix} = \mathbf{P}(1) \begin{bmatrix} b_2(1) \\ a_2(1) \end{bmatrix}, \qquad (2.14)$$

where

$$P_{11} = S_{12} - S_{11}S_{21}^{-1}S_{22} = T - \mathscr{R}\tau^{-1}R, \qquad (2.15a)$$

$$P_{12} = S_{11}S_{21}^{-1} = \mathscr{R}\tau^{-1}, \qquad (2.15b)$$

$$P_{21} = -S_{21}^{-1}S_{22} = -\tau^{-1}R, \qquad (2.15c)$$

$$P_{22} = S_{21}^{-1} = \tau^{-1}. \qquad (2.15d)$$

The cascading of n two-port networks is then described mathematically by

$$\begin{bmatrix} b_1(1) \\ a_1(1) \end{bmatrix} = \mathbf{P}(1)\mathbf{P}(2) \cdots \mathbf{P}(n) \begin{bmatrix} b_2(n) \\ a_2(n) \end{bmatrix}. \qquad (2.16)$$

This expression thus gives, by a simple matrix product, the reflected and incident waves on the left side of the first two-port network in terms of the

---

† Carlin and Giordano call this the T matrix.

matrix product, and the reflected and incident waves on the right side of the *n*th two-port network. The use of the matrix product for numerical analysis is discussed later. The matrix product formulation leads to a fundamental matrix $\mathbf{Q}(j)$ where $\mathbf{P}(j) = \mathbf{Q}^{-1}(j)$.

3. NORMAL MODE THEORY

The normal mode theory is a convenient way of studying the interaction of two modes in a physical system. It is essentially a method of reducing a set

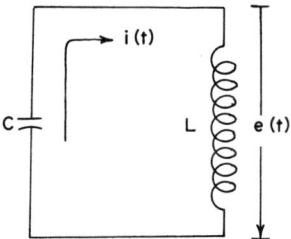

Figure 2.4. Simple oscillator.

of strongly coupled differential equations to a weakly coupled set. Approximations can then be made to find approximate solutions which are useful in engineering work.

To illustrate the normal mode theory, let us consider the simple electric circuit in Figure 2.4. Assume an initial charge q on C at $t = 0-$ such that, when C is connected to the inductor as shown, a current flows. The current $i(t)$ and the voltage are periodic functions of the frequency $\omega = (LC)^{-1/2}$. Since there is no resistance in the ideal model, the oscillations persist forever.

The equations of motion are

$$\frac{di(t)}{dt} = -\frac{e(t)}{L}, \tag{2.17a}$$

$$\frac{de(t)}{dt} = \frac{i(t)}{C}. \tag{2.17b}$$

A simple mathematical operation gives the second-order differential equation

$$\frac{d^2i(t)}{dt^2} = -\frac{1}{LC}i(t), \tag{2.18}$$

which can be solved in a straightforward manner if L and C are time independent.

2 SCATTERING MATRICES, NORMAL MODES, STATE VARIABLES

To obtain the normal mode forms for this simple circuit, let us consider a linear combination of (2.17a) and (2.17b). Multiply (2.17b) by an arbitrary function Y and add to (2.17a):

$$\frac{d}{dt}[i(t) + Ye(t)] = \frac{Y}{C}i(t) - \frac{e(t)}{L}. \tag{2.19}$$

If Y is selected to have the proper dimensions to make the product $Ye(t)$ have the dimensions of current, then Y is an admittance function. The goal of normal mode theory is to generate a wave $a(t)$, where

$$a(t) = i(t) + Ye(t) \tag{2.20}$$

such that (2.16) can be written in the forms

$$\left[\frac{d}{dt} - j\omega\right]a(t) = 0 \tag{2.21a}$$

and

$$\left[\frac{d}{dt} + j\omega\right]a^*(t) = 0. \tag{2.21b}$$

The task is then to find Y such that (2.21) holds. If (2.21) holds, then

$$\frac{d}{dt}[i(t) + Ye(t)] = j\omega[i(t) + Ye(t)], \tag{2.22}$$

where Y can only take on the values

$$Y = \pm j\omega C = \pm j\sqrt{C/L}. \tag{2.23}$$

If Y can be determined, then $a(t)$ and $a^*(t)$ are given by

$$a(t) = a(0)e^{j\omega t}, \tag{2.24a}$$

$$a^*(t) = a^*(0)e^{-j\omega t}. \tag{2.24b}$$

The two functions $a(t)$ and $a^*(t)$ can be considered as counterrotating vectors which are uncoupled. If Y can be determined, then the voltage and current can be calculated. In the case above, Y was easily determined; this is not always so.

The general approach to coupled mode theory has been made by Pierce [3]. He considered two unattenuated normal modes of vibration. Figure 2.5 illustrates a simplified picture of a two-mode system. Let a_1 and a_2 represent the complex amplitude of the two modes, each mode being properly normalized. Let $\omega_1 dt$ represent the phase shift of the a_1 mode in time dt,

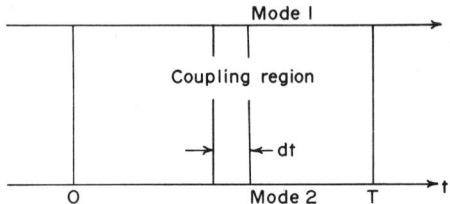

Figure 2.5. Two coupled modes.

and $\omega_2\,dt$ the phase shift of mode a_2 in the same time but both in the absence of coupling.

Since coupling is assumed, some of the energy in mode a_2 is coupled to the a_1 mode. If C_{12} specifies the coupling coefficient from mode a_1 to a_2 and if C_{21} represents the coupling from mode a_2 to a_1, then the coupled mode equations are

$$\frac{da_1}{dt} = j\omega_1 a_1 + C_{12} a_2, \qquad (2.25a)$$

$$\frac{da_2}{dt} = C_{21} a_1 + j\omega_2 a_2. \qquad (2.25b)$$

The two equations of (2.25) are of the form that is treated in this book. It is for this reason that the title of this book was selected. There are many ways that the coefficients in (2.25) can be determined; the evaluation of these coefficients is given later.

4. NORMAL MODE EQUATIONS AND THE TRANSMISSION LINE

The work which follows in this book is based on the concept of reflection and transmission of a wave on a transmission line. It therefore seems appropriate to give the derivation of the coupled differential equations of a transmission line. The equivalent circuit of a transmission line is given in Figure 2.6. The voltage drop in the length dz is given by

$$de = -\frac{\partial e}{\partial z}\,dz = R\,dz\,i + L\,dz\,\frac{\partial i}{\partial t}, \qquad (2.26a)$$

and the current change is

$$di = -\frac{\partial i}{\partial z}\,dz = G\,dz\,e + C\,dz\,\frac{\partial e}{\partial t}. \qquad (2.26b)$$

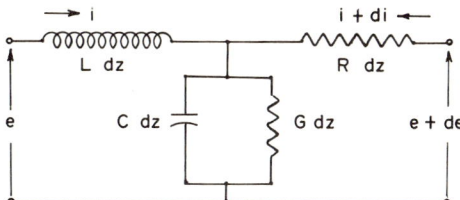

Figure 2.6 Section of transmission line.

To simplify the normal mode derivation, let $G = R = 0$, the case of a lossless line. It is then easy to show that, if

$$b_\pm(z, t) = \frac{1}{2\sqrt{Z_0}} [e(z, t) \pm Z_0(z, t)i(z, t)] \tag{2.27}$$

and

$$Z_0(z, t) = \pm\sqrt{L/C},$$

then

$$\frac{\partial b_\pm}{\partial z} = \pm\sqrt{LC}\,\frac{\partial b_\pm}{\partial t}. \tag{2.28}$$

It is interesting to note that (2.27) is the definition of an incident and a reflected wave as defined by the scattering matrix; see (2.6a) and (2.6b). Thus the ratio $b_-(z, t)/b_+(z, t)$ is the reflection coefficient and $[b_+(z, t) - b_-(z, t)]/b_+(z, t)$ is the transmission coefficient.

If $b_\pm(z, t)$ is defined as a combination of $a_\pm(z)$ and $a_\pm^*(z)$, that is,

$$b_\pm(z, t) = a_\pm(z)e^{j\omega t} + a_\pm^*(z)e^{-j\omega t}, \tag{2.29}$$

the normal mode equations follow directly:

$$\left[\frac{d}{dz} \pm j\beta\right]a_\pm(z) = 0, \tag{2.30a}$$

$$\left[\frac{d}{dz} \mp j\beta\right]a_\pm^*(z) = 0. \tag{2.30b}$$

The function β is equal to $\omega\sqrt{LC}$.

Since $b_-(z, t)/b_+(z, t)$ is the reflection coefficient, it can be shown that this ratio obeys a Riccati differential equation. The normal mode theory avoids the Riccati equation by considering the uncoupled equations (2.30a) and (2.30b). It is not always possible to uncouple the equations for $a(z)$ and $a^*(z)$, and the system of equations in (2.25) must be solved. The objective in the following work is to solve such equations, and the method is applicable to the normal mode equations.

5. STATE VARIABLES AND STATE SPACE

Several books have recently appeared in print whereby control systems are described in terms of state functions. The state-variable representation results in a system description in terms of n first-order differential equations. The use of n first-order differential equations to describe the dynamics of a control system is a different approach from the conventional transfer function manipulation which has been a standard method of analysis in the past. The coupled mode equations described herein are first-order differential equations and applicable to state variables. The transformation of the state variable equations into coupled mode equations is given here.

The following discussion is only an introduction to state variables, and it indicates that the coupled mode equations treated in this book are of the same form. The reader interested in state variables will find the numerical scheme given useful in solving general state space problems.

Assume that a system is characterized in state space by the equations of (1.3) and (1.4)

$$\dot{\mathbf{X}} = \mathbf{AX} + \mathbf{BZ} \tag{1.3}$$

$$\mathbf{Y} = \mathbf{CX} + \mathbf{DZ} \tag{1.4}$$

In general, **A**, **B**, **C**, and **D** are time varying matrices and **X** and **Z** are column vectors. If **Z** is the excitation vector for the system, **X** denotes the states of the system and **Y** is the output. The matrix **A** describes the system in an unforced state, **B** describes the coupling between the input **Z** and the system, **C** and **D** are coupling matrices which describe how the system states and the input are coupled to the output. A block diagram for equations (1.3) and (1.4) is shown in Figure 2.7.

If the matrix **A** of equation (1.3) is a constant, the solution can be found in a rather straightforward manner by using Laplace transforms. Consider the homogeneous part of (1.3) and take the Laplace transform of the equation:

$$\mathbf{X}(s) - \mathbf{X}(0) = \mathbf{AX}(s); \tag{2.31}$$

then

$$\mathbf{X}(s) = [s\mathbf{I} - \mathbf{A}]^{-1}\mathbf{X}(0). \tag{2.32}$$

The inverse transform can now be found and the homogeneous solution is

$$\mathbf{X}(t) = \mathbf{\Omega}(t)\mathbf{X}(0), \tag{2.33}$$

where

$$\mathbf{\Omega}(t) = \mathscr{L}^{-1}\{[s\mathbf{I} - \mathbf{A}]^{-1}\}. \tag{2.34}$$

2 SCATTERING MATRICES, NORMAL MODES, STATE VARIABLES

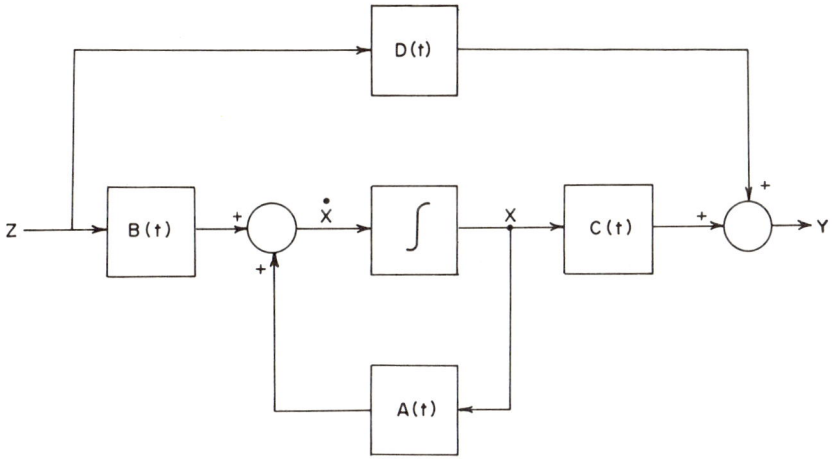

Figure 2.7. Block diagram for state equations.

The complete solution for (1.3) is then found from the equation

$$\mathbf{X}(t) = \mathbf{\Phi}(t)\mathbf{X}(0) + \int_0^t \mathbf{\Phi}(t-\tau)\mathbf{B}\mathbf{Z}(\tau)\,d\tau \tag{2.35}$$

and the output is then given by

$$\mathbf{Y}(t) = \mathbf{C}\left\{\mathbf{\Phi}(t)\mathbf{X}(0) + \int_0^t \mathbf{\Phi}(t-\tau)\mathbf{B}\mathbf{Z}(\tau)\,d\tau\right\} + \mathbf{D}\mathbf{Z}(t). \tag{2.36}$$

The procedure above is simple to carry out if $\mathbf{A}$ is a constant. If $\mathbf{A}$ is a time varying function, as it may be, the transform method cannot be applied. The transition matrix $\mathbf{\Phi}(t)$ must then be obtained by some other means.

In the work that follows, invariant imbedding is utilized to find $\mathbf{\Phi}(t)$. The fundamental matrix, which is found by the numerical method, is designated by $\mathbf{Q}(0, t)$; this matrix is nothing more than the state transition matrix. The method of obtaining $\mathbf{Q}(0, t)$ by recursive relations and the means of carrying out the indicated integration in a recursive fashion are described.

Consider again equations (1.1), (1.2), and (1.3),

$$\frac{du}{dx} = Bu(x) + Av(x) + e(x), \tag{1.1}$$

$$-\frac{dV}{dx} = Du(x) + Cv(x) + f(x), \tag{1.2}$$

$$\dot{\mathbf{X}} = \mathbf{A}\mathbf{X} + \mathbf{B}\mathbf{Z}, \tag{1.3}$$

where (1.1) and (1.2) are in the spatial domain and (1.3) is in the time domain. Assume that **A** and **B** are $n \times n$ matrices which are of even order. To obtain the coupled equations of (1.1) and (1.2), partition **X** into two sets of equations of order $n/2$. Let the first set be denoted by $\dot{\mathbf{X}}_1(t)$ and the second by $\dot{\mathbf{X}}_2(t)$. Then, changing the variable x to t and setting $X_1(t) = u(x)$ and $X_2(t) = v(x)$, the desired form is obtained. The numerical procedure to be described then holds.

If **A** and **B** are $n \times n$ matrices of odd order, one additional equation can be generated to complete the set:

$$\dot{x}_{n+1}(t) = -x_{n+1}(t).$$

The matrices **A** and **B** then have the correct order necessary for partitioning into two sets of the same order. There are perhaps other ways of circumventing this difficulty, but this procedure works.

The books by DeRusso, Roy, and Close [4] and Shultz and Melsa [5] are excellent texts for an introduction to state space. The comprehensive text by Zadeh and Desoer [6] establishes the mathematical rigor for state space and should be consulted for the necessary proofs, etc.

REFERENCES

1. H. Carlin and A. Giordano, *Network Theory: An Introduction to Reciprocal and Nonreciprocal Circuits*, Prentice-Hall, Inc., Englewood Cliffs, N.J., 1964.
2. W. Louisell, *Coupled Modes and Parametric Electronics*, John Wiley and Sons, New York, 1960.
3. J. R. Pierce, Coupling of Modes of Propagation, *J. Appl. Phys.*, 25 (1964), 179.
4. P. M. DeRusso, R. J. Roy, and C. M. Close, *State Variables for Engineers*, John Wiley and Sons, 1966.
5. D. G. Schultz and J. L. Melsa, *State Functions and Linear Control Systems*, McGraw-Hill Book Co., New York, 1963.
6. L. Zadeh and C. Desoer, *Linear Systems Theory: The State Space Approach*, McGraw-Hill Book Co., New York, 1963.

CHAPTER 3

Recursive Equations

Invariant imbedding has been used in the past to solve problems in wave propagation (Adams and Denman [1]), neutron diffusion (Bellman, Kalaba, and Wing [2]), wave branching (Bellman and Kalaba [3]), gamma-ray transmission and reflection (Shimizu and Mizuta [4]), and in numerous other problem areas. The purpose of the work here is to extend the method to the multimode problem. Many of the ideas discussed in this book are simple extensions of the mathematical work of Redheffer [5] and Reid [6]. The major deviation from their work is the almost exclusive use of the recursive relations rather than the matrix differential equations.

The term *multimode* is used to designate any process in which an exchange of energy exists between competing phenomena. The energy exchange may be between the currents in the loops of an electrical network or between electromagnetic waves in a coupling medium.

Although most of the work is concerned with establishing the method and analyzing wave propagation, the method will be seen to be more general. If a physical process is defined by the coupled differential equations

$$\frac{du(x)}{dx} = B(x)u(x) + A(x)v(x), \tag{3.1a}$$

$$-\frac{dv(x)}{dx} = D(x)u(x) + C(x)v(x), \tag{3.1b}$$

with $A(x)$ and $D(x)$ nonzero, then coupling occurs between $u(x)$ and $v(x)$. The numerical solution of (3.1) is discussed in the next chapter. The recursive equations derived in this chapter are an integral part of the numerical technique for solving (3.1).

I. THE SINGLE-MODE PROBLEM

To introduce the concept of invariant imbedding, the simple problem of a single propagating mode is studied first. Let us consider an electromagnetic wave propagating in an inhomogeneous medium. The wave is continuously reflected along its path and only a certain amount of the energy can be transmitted through the medium. The propagation of such a wave is analogous to a wave on a transmission line with a variable impedance along the line. Energy may be transmitted along the line but the changing impedance will excite reflected waves. The assumption is made that superposition holds for all waves on the transmission line; that is, a wave at frequency ω_1 does not couple any of its energy to a wave of frequency ω_2.

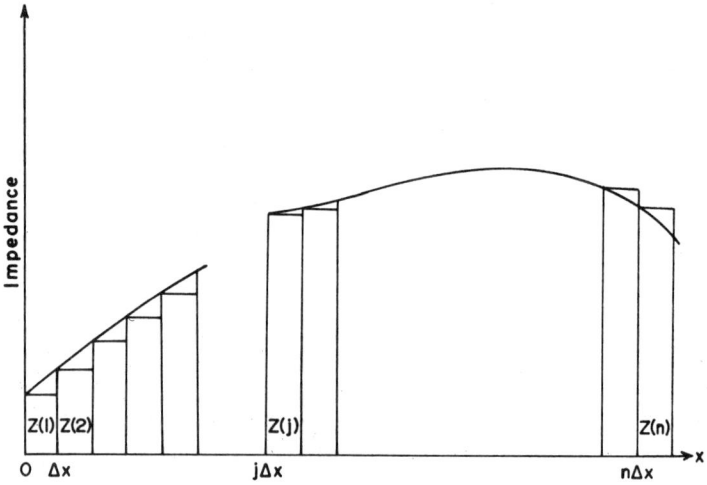

Figure 3.1. Transmission line of discrete sections.

Assume that the transmission line is of length a and that the impedance is a function of x. Let the line be broken into discrete sections of length Δx, each section having a constant impedance $Z(j)$, where j signifies the jth section of length Δx. If the line is of length a, then there are n sections of length Δx with impedance $Z(j)$ for $j = 1, 2, 3, \ldots, n$.

Let the impedance $Z(j)$ of the line vary as shown in Figure 3.1. Consider the jth section of the line as well as the j and $j + 1$ interfaces, the interfaces at $(j - 1)\Delta x$ and $j\Delta x$. Each section of the line has a constant impedance and there exists an abrupt change of impedance at each interface. A wave incident on the jth interface from the left undergoes a reflection at that interface. Part of the wave is transmitted through the interface into the section of impedance $Z(j)$, where it undergoes a phase shift before striking the

3 RECURSIVE EQUATIONS

$(j + 1)$th interface. The wave striking the $(j + 1)$th interface is partially reflected and transmitted. Each of the resulting waves strikes other interfaces and multiple reflections occur along the line at each interface. The calculation of the effective reflection coefficient for the entire medium is quite involved if all interfaces are considered simultaneously. Invariant imbedding makes it possible to consider a single section and the two bounding interfaces in a collective fashion. The effective transmission matrix is calculated for the first section; this information is then used to find two sections in cascade. This process is continued until all sections of the line have been cascaded together.

To illustrate the method, consider first the case of the interface at $(j - 1)\Delta x$ with a semi-infinite medium on both sides of the interface. The impedance on the left of the interface is $Z(j - 1)$, that on the right is $Z(j)$. Let a wave be incident on both sides of the interface, where an odd subscript is used to denote a wave moving to the right, and an even subscript a wave moving to the left. If E_{i1} is the wave incident on the left side of the interface and E_{i2} the wave incident on the right side (see Figure 3.2), the wave moving away from the interface to the right consists of a reflected and transmitted wave. A similar wave exists on the left side of the interface and it moves toward the left. The transmitted fields are given by

$$E_{t1}(j) = \tau(j)E_{i1}(j - 1) + R(j)E_{i2}(j), \tag{3.2a}$$

$$E_{t2}(j - 1) = \mathcal{R}(j)E_{i1}(j - 1) + T(j)E_{i2}(j), \tag{3.2b}$$

where $E_{i1}(j - 1)$ is the field within the $(j - 1)$th slab and $E_{i2}(j)$ denotes the field in the jth slab. The jth interface is the only one under consideration, and the reflection coefficients $R(j)$ and $R(j)$ and the transmission coefficients $T(j)$ and $\tau(j)$ are for the simple interface. The matrix

$$\mathbf{T}(j) = \begin{bmatrix} \tau(j) & R(j) \\ \mathcal{R}(j) & T(j) \end{bmatrix} \tag{3.3}$$

is hereafter called the transmission matrix.

The interface of Figure 3.2 has a semi-infinite medium on each side and the four coefficients of (3.3) are easily found. Only primary reflections occur and secondary reflections need not be considered. To find the four coefficients, the imposed boundary conditions of continuity of E and H are used. The four

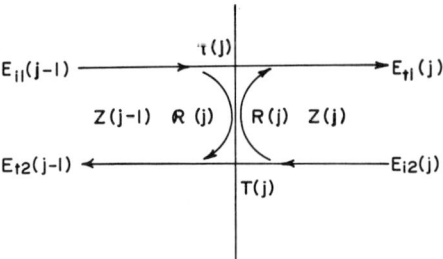

Figure 3.2. A simple interface.

coefficients are as given below.

$$R(j) = \frac{Z(j-1) - Z(j)}{Z(j-1) + Z(j)}, \tag{3.3a}$$

$$\mathscr{R}(j) = \frac{Z(j) - Z(j-1)}{Z(j-1) + Z(j)}, \tag{3.3b}$$

$$T(j) = \frac{2Z(j-1)}{Z(j-1) + Z(j)}, \tag{3.3c}$$

$$\tau(j) = \frac{2Z(j)}{Z(j-1) + Z(j)}. \tag{3.3d}$$

Knowledge of $Z(x)$ makes it possible to compute each of the four coefficients. Although the impedance concept has been used in (3.3), the coefficients can be defined in terms of the refractive index of the medium.

To simplify the algebra of generating the recursive equations for the cascading process, the signal flow graph is used to find the cascaded transmission matrix. The signal flow graph not only simplifies the mathematics but also gives a visual picture of the scattering process. The recursive equations of the cascaded line can be derived by inspection of the graphs.

Each interface can be described by a signal flow graph, the cascaded interfaces can be represented by a combination of simple graphs. The simple interface is illustrated by a signal flow graph in Figure 3.3; it describes mathematically the scattering process. The signal flow graph for the simple interface involves the four coefficients as defined in (3.3). Each interface that is cascaded to one or more interfaces is characterized by the four coefficients of that particular interface. The cascaded flow graph must also include the phase shift between adjacent interfaces. Mason's gain rule can be employed to find the cascaded coefficients directly from the graph.

Suppose now that the $(j + 1)$th interface is cascaded to the right of the

3 RECURSIVE EQUATIONS

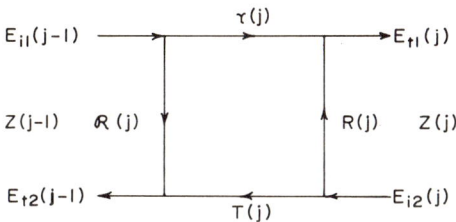

Figure 3.3. Flow graph for simple interface.

jth interface. The semi-infinite medium $Z(j-1)$ is still present on the left of the graph and a semi-infinite medium of impedance $Z(j+1)$ appears on the right. The two interfaces are separated by a finite medium of thickness Δx and its characteristic impedance is $Z(j)$. The configuration is shown in Figure 3.4.

The field equations for the jth interface are still valid, but a second set for the $(j+1)$th interface must now be added to describe completely the scattering process. The interface equations for the $(j+1)$th interface are

$$E_{t1}(j+1) = \tau(j+1)E_{i1}(j) + R(j+1)E_{i2}(j+1), \quad (3.4a)$$

$$E_{t2}(j+1) = \mathscr{R}(j+1)E_{i1}(j) + T(j+1)E_{i2}(j+1), \quad (3.4b)$$

and the phase shift in the medium $Z(j)$ is such that

$$E_{i1}(j+1) = E_{t1}(j)\exp[-iK(j)\Delta x], \quad (3.5a)$$

$$E_{i2}(j) = E_{t2}(j+1)\exp[-iK(j)\Delta x]. \quad (3.5b)$$

where $K(j)$ is the wave number of the medium with impedance $Z(j)$.

The signal flow graph for the two-interface single-slab models can now be constructed from (3.2), (3.4), and (3.5). The signal flow graph for the two waves is shown in Figure 3.5. The field equations for the model of Figure 3.5

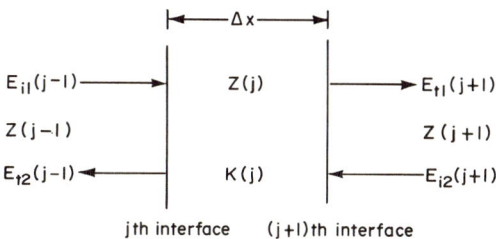

Figure 3.4. Single-slab model.

can now be written as given below.

$$E_{t1}(j+1) = \tau(j, j+1)E_{i1}(j-1) + R(j, j+1)E_{i2}(j+1), \quad (3.6a)$$
$$E_{t2}(j-1) = \mathscr{R}(j, j+1)E_{i1}(j-1) + T(j, j+1)E_{i2}(j+1). \quad (3.6b)$$

The coefficients $\tau(j, j+1)$, $\mathscr{R}(j, j+1)$, $R(j, j+1)$, and $T(j, j+1)$ are now defined as the composite coefficients. The composite coefficients are the mathematical equivalent of the coefficients for the two interfaces in cascade, taking into account multiple reflections.

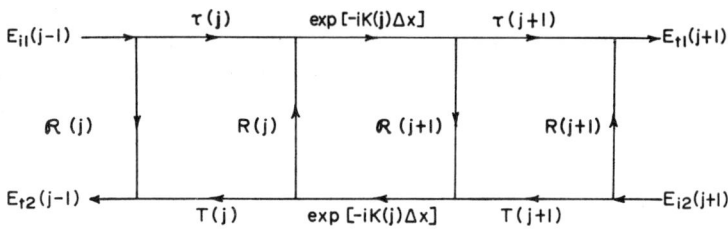

Figure 3.5. Signal flow graph for single slab.

The composite coefficients are quite easy to find from the flow graph given in Figure 3.5. Note that, if $E_{i2}(j+1)$ is zero in (3.6a), then $\tau(j, j+1)$ is the ratio of the output $E_{t1}(j+1)$ to the input $E_{i1}(j-1)$. This ratio can be found by applying Mason's gain rule directly to Figure 3.5. The other composite coefficients are found by selectively making the other inputs zero and solving for the associated composite coefficients:

$$\tau(j, j+1) = \frac{\tau(j)\tau(j+1)\exp[-iK(j)\Delta x]}{[1 - R(j)\mathscr{R}(j+1)]\exp[-2iK(j)\Delta x]}, \quad (3.7a)$$

$$T(j, j+1) = \frac{T(j)T(j+1)\exp[-iK(j)\Delta x]}{[1 - R(j)\mathscr{R}(j+1)]\exp[-2iK(j)\Delta x]}, \quad (3.7b)$$

$$R(j, j+1) = R(j+1) + \frac{\tau(j+1)T(j+1)R(j)\exp[-2iK(j)\Delta x]}{[1 - R(j)\mathscr{R}(j+1)]\exp[-2iK(j)\Delta x]}, \quad (3.7c)$$

$$\mathscr{R}(j, j+1) = \mathscr{R}(j) + \frac{\tau(j)T(j)\mathscr{R}(j+1)\exp[-2iK(j)\Delta x]}{[1 - R(j)\mathscr{R}(j+1)]\exp[-2iK(j)\Delta x]}, \quad (3.7d)$$

Coefficients for a single interface are designated by a single letter within the bracket of the coefficient. The right side of each of the four equations of (3.7) depends on the simple interface coefficients. The composite coefficients,

3 RECURSIVE EQUATIONS

$\tau(j, j + 1)$, $R(j, j + 1)$, $\mathcal{R}(j, j + 1)$, and $T(j, j + 1)$, are thus determined from the properties of the single interfaces. The interface coefficients are determined from (3.3), where it is assumed that $Z(x)$ is known everywhere.

Having disposed of the simple interface and the single-slab models, consider now the multiple-slab problem. Assume that a total of j slabs are to be cascaded together. Suppose that slabs 1, 2, 3, ..., $j - 1$ have been cascaded together and that the four composite coefficients are known. The problem is then to find the composite coefficients for the slabs 1, 2, 3, ..., $j - 1$, j in cascade. To find the composite coefficients for the j slabs in cascade, note that there are now $(j + 1)$ interfaces. A simple change in (3.7) is all that is required to calculate the jth slab composite coefficients. If the composite coefficients for the first $(j - 1)$ slabs are known, then replace $\tau(j), R(j), T(j)$, and $\mathcal{R}(j)$ in (3.7) by $\tau(1, 2, 3, \ldots, j), R(1, 2, 3, \ldots, j), T(1, 2, 3, \ldots, j)$, and $\mathcal{R}(1, 2, 3, \ldots, j)$, respectively. The composite coefficients $\tau(1, 2, 3, \ldots, j, j + 1)$, $R(1, 2, 3, \ldots, j, j + 1)$, $T(1, 2, 3, \ldots, j, j + 1)$, and $\mathcal{R}(1, 2, 3, \ldots, j, j + 1)$ are then given in (3.8), where the argument for the coefficients indicates the number of interfaces.

$$\tau(1, 2, 3, \ldots, j, j + 1) = \frac{\tau(1, 2, 3, \ldots, j)\tau(j + 1)\exp[-iK(j)\Delta x]}{[1 - R(1, 2, 3, \ldots, j)\mathcal{R}(j + 1)]\exp[-2iK(j)\Delta x]}, \quad (3.8a)$$

$$R(1, 2, 3, \ldots, j, j + 1) = R(j + 1) + \frac{\tau(j + 1)T(j + 1)R(1, 2, 3, \ldots, j)\exp[-2iK(j)\Delta x]}{[1 - R(1, 2, 3, \ldots, j)\mathcal{R}(j + 1)]\exp[-2iK(j)\Delta x]}, \quad (3.8b)$$

$$T(1, 2, 3, \ldots, j, j + 1) = \frac{T(1, 2, 3, \ldots, j)T(j + 1)\exp[-iK(j)\Delta x]}{[1 - R(1, 2, 3, \ldots, j)\mathcal{R}(j + 1)]\exp[-2iK(j)\Delta x]}, \quad (3.8c)$$

$$\mathcal{R}(1, 2, 3, \ldots, j, j + 1) = \mathcal{R}(1, 2, 3, \ldots, j)$$
$$+ \frac{\tau(1, 2, 3, \ldots, j)T(1, 2, 3, \ldots, j)\mathcal{R}(j + 1)\exp[-2iK(j)\Delta x]}{[1 - R(1, 2, 3, \ldots, j)\mathcal{R}(j + 1)]\exp[-2iK(j)\Delta x]}. \quad (3.8d)$$

The composite coefficients defined by (3.8) are calculated by a recursive process. If there is no interface, then all of the wave is transmitted and no reflections occur.

$$R(0) = \mathcal{R}(0) = 0,$$
$$T(0) = \tau(0) = 1.$$

Therefore we start with cascading interfaces 1 and 2 together, using the composite coefficients for the two interfaces to cascade 1, 2, and 3 together. This process is continued until all slabs have been included.

The four equations of (3.8) are ideal for computational use on a high speed digital computer. Although the equations for the four coefficients are given in (3.8), it may not always be necessary to use all four. If there is only an incident wave on the right side of the medium, (3.8b) and (3.8c) can be used to find the reflection and transmission coefficients for the entire medium. Since there is no wave incident on the left, the field on both sides of the medium is fully characterized. Care should be taken in selecting the step size because artificial "resonances" may result (see Adams and Denman [7]) when this model is used.

2. THE MULTIMODE PROBLEM

The single-mode problem is relatively simple since only scalar quantities are involved in the composite coefficients. Whenever two or more modes propagate along the structure, the modeling becomes more complex if interaction is allowed between the modes or waves. Not only must the transmission and reflection phenomena be included in describing the wave behavior, the coupling between modes must also be taken into account. For example, if two modes are considered, the number of reflection and transmission coefficients increases to eight, and eight coupling coefficients must be introduced. The number of composite coefficients which must be defined is sixteen. By properly defining the field quantities the coefficients can be written as matrices; consequently only four 2×2 matrix coefficients need to be introduced. These four matrix coefficients are sufficient to handle any number of interacting modes, the matrices being of size $m \times m$ for a m-mode problem.

Consider a structure in which two modes are present and the structure allows energy to be exchanged between modes. Such a structure is illustrated in Figure 3.6 where two waves are incident on each side of the structure. It is assumed that the structure indicated by the block is an interface where coupling occurs. If coupling occurs, then $E_{t1}(j)$ consists of contribution from four sources. Assume that the waves with subscripts $i1$, $t1$, $i2$, and $t2$ are similar and the waves with subscripts $i3$, $t3$, $i4$, and $t4$ are also similar. $E_{t1}(j)$ is made up of transmitted and reflected components of like waves and components from the dissimilar wave being coupled into $E_{t1}(j)$. The reflection and transmission coefficients are denoted in the usual manner, but subscripts are assigned to $T(j)$, $\tau(j)$, $R(j)$, and $\mathscr{R}(j)$.

3 RECURSIVE EQUATIONS

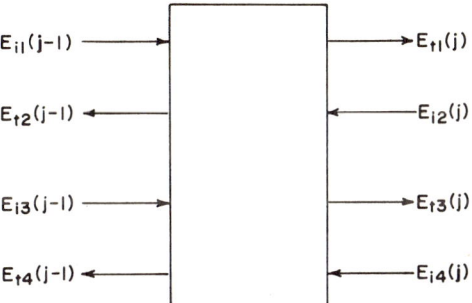

Figure 3.6. Coupled mode model.

The transmitted waves can now be described mathematically for the structure of Figure 3.6.

$$\begin{aligned}
E_{t1}(j) &= \tau_{11}(j)E_{i1}(j-1) + \tau_{13}(j)E_{i3}(j-1) \\
&\quad + R_{12}(j)E_{i2}(j) + R_{14}(j)E_{i4}(j), \\
E_{t2}(j-1) &= \mathscr{R}_{21}(j)E_{i1}(j-1) + \mathscr{R}_{23}(j)E_{i3}(j-1) \\
&\quad + T_{22}(j)E_{i2}(j) + T_{24}(j)E_{i4}(j), \\
E_{t3}(j) &= \tau_{31}(j)E_{i1}(j-1) + \tau_{33}(j)E_{i3}(j-1) \\
&\quad + R_{32}(j)E_{i2}(j) + R_{34}(j)E_{i4}(j), \\
E_{t4}(j-1) &= \mathscr{R}_{41}(j)E_{i1}(j-1) + \mathscr{R}_{43}(j)E_{i3}(j-1) \\
&\quad + T_{42}(j)E_{i2}(j) + T_{44}(j)E_{i4}(j).
\end{aligned} \quad (3.9)$$

To simplify (3.9), consider the incident and transmitted components on each side of the structure. The components can be considered as vectors; let

$$u(j-1) = \begin{bmatrix} E_{i1}(j-1) \\ E_{i3}(j-1) \end{bmatrix}, \quad u(j) = \begin{bmatrix} E_{t1}(j) \\ E_{t3}(j) \end{bmatrix}, \quad (3.10a)$$

and

$$v(j-1) = \begin{bmatrix} E_{t2}(j-1) \\ E_{t4}(j-1) \end{bmatrix}, \quad v(j) = \begin{bmatrix} E_{i2}(j) \\ E_{i4}(j) \end{bmatrix}. \quad (3.10b)$$

Rearranging (3.9) and substituting (3.10) into (3.9) yields

$$\begin{bmatrix} u(j) \\ v(j-1) \end{bmatrix} = \begin{bmatrix} \tau(j) & R(j) \\ \mathscr{R}(j) & T(j) \end{bmatrix} \begin{bmatrix} u(j-1) \\ v(j) \end{bmatrix}, \quad (3.11)$$

where $\tau(j)$, $R(j)$, $\mathscr{R}(j)$, and $T(j)$ are 2×2 matrices. The diagonal terms of

$\tau(j)$ and $T(j)$ are transmission coefficients, whereas the off-diagonal terms are coupling coefficients; $R(j)$ and $\mathcal{R}(j)$ have reflection coefficients along their diagonals with coupling terms off the diagonal. Although only two modes were considered in (3.9), (3.11) is valid for any number of modes. The matrices $R(j)$, $T(j)$, etc., must be enlarged to account for all modes.

To cascade the simple interface of Figure 3.6 to a second interface, the phase shift between the two interfaces must be included. Since the waves are now defined through matrix equations, the phase shift in the homogeneous slab separating the interfaces must be specified as a matrix. Assume that no coupling occurs in the slab, that only a phase shift occurs. The phase shift

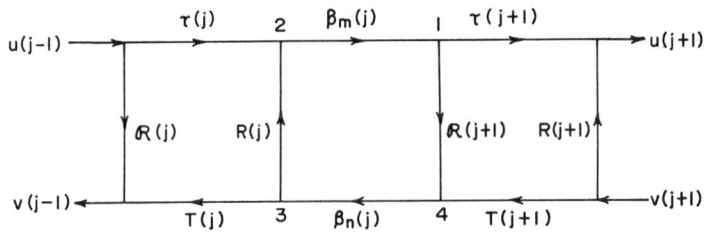

Figure 3.7. Flow graph for single-slab coupled model.

matrix must be of the same order as $T(j)$, $\tau(j)$, $R(j)$, and $\mathcal{R}(j)$. Therefore, for the waves traveling from left to right,

$$\beta_m(j) = \begin{bmatrix} \exp[-iK_1(j)\Delta x] & 0 \\ 0 & \exp[-iK_3(j)\Delta x] \end{bmatrix}, \quad (3.12a)$$

and, for those going in the opposite direction,

$$\beta_n(j) = \begin{bmatrix} \exp[-iK_2(j)\Delta x] & 0 \\ 0 & \exp[-iK_4(j)\Delta x] \end{bmatrix}. \quad (3.12b)$$

In some problems $\beta_m(j) = \beta_n(j)$, but care should be taken in making this assumption since not all media obey this equality.

The signal flow graph for (3.11) is identical in form with that of the scalar problem for which the flow graph is given in Figure 3.3. The transmittance of the flow graphs for (3.11) is expressed in matrix rather than scalar forms as given for the single-mode problem. Incident and transmitted waves are now treated as vectors, defined in (3.10). The signal flow graph for the j and $j+1$ interfaces cascaded together is now given in Figure 3.7.

The composite matrices $\tau(j, j+1)$, $T(j, j+1)$, $R(j, j+1)$ and

3 RECURSIVE EQUATIONS

$\mathscr{R}(j, j+1)$ can be found from Figure 3.7, although Mason's gain formula is no longer valid. The transmittance of the graph is in matrix form, and the order of the matrices must be preserved. A modified gain formula can be devised which is valid for the graph of Figure 3.7, but it is not general and should be used with care. To find the composite coefficient $\tau(j, j+1)$, locate all forward paths between $u(j-1)$ and $u(j+1)$, as well as all closed loops of the graph, in this case the loop formed by nodes 1, 2, 3, and 4. The forward path between $u(j-1)$ and $u(j+1)$ must be traversed in the reverse direction. Each transmittance encountered along the reversed path becomes a postmultiplier for the matrix path. If a loop is encountered, the matrix I minus the loop gain must be found, the loop gain being formed also in the reverse direction. The inverse of I minus the loop gain also becomes a postmultiplier and, after the loop has been taken into account, continue along the remainder of the reverse path.

If the composite coefficient $\tau(j, j+1)$ is sought, the reverse path is along $\tau(j+1)$, $\beta_m(j)$, and $\tau(j)$. A loop is encountered at node 1; the matrix $[I - \beta_m(j)R(j)\beta_n(j)\mathscr{R}(j+1)]$ accounts for the loop. The composite coefficient $\tau(j, j+1)$ is then

$$\tau(j, j+1) = \tau(j+1)[I - \beta_m(j)R(j)\beta_n(j)\mathscr{R}(j+1)]^{-1}\beta_m(j)\tau(j). \tag{3.13a}$$

Note that the transversal order has been preserved in (3.13a).

The remaining composite coefficients for the matrix flow graph are

$$T(j, j+1) = T(j)[I - \beta_n(j)\mathscr{R}(j+1)\beta_m(j)R(j)]^{-1}\beta_n(j)T(j+1), \tag{3.13b}$$

$$\mathscr{R}(j, j+1) = T(j)[I - \beta_n(j)\mathscr{R}(j+1)\beta_m(j)R(j)]^{-1}\beta_n(j)\mathscr{R}(j+1)$$
$$\cdot \beta_m(j)\tau(j) + \mathscr{R}(j), \tag{3.13c}$$

$$R(j, j+1) = \tau(j+1)[I - \beta_m(j)R(j)\beta_n(j)\mathscr{R}(j+1)]^{-1}\beta_m(j)R(j)$$
$$\cdot \beta_n(j)T(j+1) + R(j+1). \tag{3.13d}$$

The composite coefficients in (3.13) can be used to calculate the effective transmission, reflection, and coupling coefficients for interacting waves in the two-interface problem. The composite coefficients depend on the interface coefficients for the jth and the $(j+1)$th interfaces, as well as the phase shifts in the slab.

To cascade many sections requires that (3.13) be modified as in the single-mode problem. Let $T(1, 2, 3, \ldots, j)$, $\tau(1, 2, 3, \ldots, j)$, $R(1, 2, 3, \ldots, j)$, and $\mathscr{R}(1, 2, 3, \ldots, j)$ denote the first j sections in cascade. To add the interface for the $j+1$ section, replace $T(j)$, $\tau(j)$, $R(j)$, and $\mathscr{R}(j)$ in (3.13) by

$T(1, 2, 3, \ldots, j)$, $\tau(1, 2, 3, \ldots, j)$, $R(1, 2, 3, \ldots, j)$, and $\mathcal{R}(1, 2, 3, \ldots, j)$, respectively. The composite coefficients for $(j + 1)$ interfaces in cascade are

$$\tau(1, 2, 3, \ldots, j, j + 1) = \tau(j + 1)[I - \beta_m(j)R(1, 2, 3, \ldots, j)$$
$$\cdot \beta_n(j)\mathcal{R}(j + 1)]^{-1}\beta_m(j)\tau(1, 2, 3, \ldots, j), \quad (3.14a)$$

$$T(1, 2, 3, \ldots, j, j + 1) = T(1, 2, 3, \ldots, j)[I - \beta_m(j)\mathcal{R}(j + 1)\beta_m(j)$$
$$\cdot R(1, 2, 3, \ldots, j)]^{-1}\beta_n(j)T(j + 1), \quad (3.14b)$$

$$\mathcal{R}(1, 2, 3, \ldots, j, j + 1) = T(1, 2, 3, \ldots, j)[I - \beta_n(j)\mathcal{R}(j + 1)\beta_m(j)$$
$$\cdot R(1, 2, 3, \ldots, j)]^{-1}\beta_n(j)\mathcal{R}(j + 1)$$
$$\cdot \beta_m(j)\tau(1, 2, 3, \ldots, j) + \mathcal{R}(1, 2, 3, \ldots, j), \quad (3.14c)$$

$$R(1, 2, 3, \ldots, j, j + 1) = \tau(j + 1)[I - \beta_m(j)R(1, 2, 3, \ldots, j)\beta_n(j)$$
$$\cdot \mathcal{R}(j + 1)]^{-1}\beta_m(j)R(1, 2, 3, \ldots, j)$$
$$\cdot \beta_n(j)T(j + 1) + R(j + 1). \quad (3.14d)$$

The composite coefficients are utilized in the same manner as the scalar equations. First calculate the interface constants $\tau(1)$, $T(1)$, $R(1)$, and $R(1)$. Since $T(0) = \tau(0) = I$ and $R(0) = \mathcal{R}(0) = 0$, the first interface need not be cascaded. The matrices $\tau(1)$, $T(1)$, $R(1)$, and $\mathcal{R}(1)$ are then used with $\tau(2)$, $T(2)$, $R(2)$ and $\mathcal{R}(2)$ to find $\tau(1, 2)$ and the other composite coefficients. The description of the method by which the interface constants are found is postponed until later.

3. THE RECURSIVE EQUATIONS AND THE CONTINUOUS MEDIUM

The preceding sections have been concerned with the derivation of the transmission, reflection, and coupling coefficients for a discrete version of the transmission line. One of the difficulties with the discrete case is that artificial "resonances" are excited and can lead to erroneous answers. A smoothing of these resonances can be obtained by treating the problem in a continuous fashion. Recursive equations can be derived for the continuous case; the discrete case is an approximation to the continuous case where $\Delta x \to 0$ in the limit. The interface constants are determined directly from the differential equations for the continuous medium. As will be seen in the next chapter, the continuous formulation is important in handling coupled differential equations.

3 RECURSIVE EQUATIONS

Assume that the transmission line has been described by differential equations and that $u(x)$ and $v(x)$ are known at x. The variables $u(x)$ and $v(x)$ are treated from this point on as vector functions. A slab of thickness Δx is now to be added to the transmission line, and the functions $u(x + \Delta x)$ and $v(x + \Delta x)$ are sought. The transmission line is depicted in Figure 3.8. The field variables $u(x)$ and $v(x)$ for the length x are specified in terms of $u(0)$ and $v(0)$ as below.

$$u(x) = \tau(0, x)u(0) + R(0, x)v(x), \quad (3.15a)$$

$$v(0) = \mathscr{R}(0, x)u(0) + T(0, x)v(x). \quad (3.15b)$$

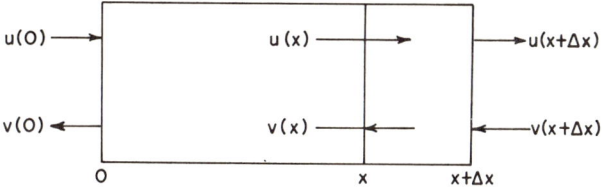

Figure 3.8. Continuous line.

Since the transmission line is continuous, the field equations for the line between x and $x + \Delta x$ depend on $u(x)$ and $v(x)$:

$$u(x + \Delta x) = \tau(x, x + \Delta x)u(x) + R(x, x + \Delta x)v(x + \Delta x), \quad (3.16a)$$

$$v(x) = \mathscr{R}(x, x + \Delta x)u(x) + T(x, x + \Delta x)v(x + \Delta x). \quad (3.16b)$$

The coefficients

$$\tau(x, x + \Delta x), R(x, x + \Delta x), \mathscr{R}(x, x + \Delta x), \text{ and } T(x, x + \Delta x)$$

are the effective coefficients for the thickness Δx. As defined earlier, these coefficients are the incremental coefficients. The four coefficients of (3.15), $\tau(0, x)$, $R(0, x)$, $\mathscr{R}(0, x)$, and $T(0, x)$ are the composite coefficients for the medium of thickness x.

The field variables $u(x)$ and $v(x)$ can be eliminated from (3.15) and (3.16) and the composite coefficients found. The flow graph for the continuous case is similar to the graph of Figure 3.7, the major difference being that the discrete coefficients are replaced by the continuous ones. Because of the similarity, the continuous graph is not given. The phase shifts are no longer present, as the incremental coefficients now include the phase change. The recursive equations for the continuous medium are written directly from (3.14) where $\beta_m(j) = \beta_n(j) = I$, and continuous coefficients are utilized.

$$\tau(0, x + \Delta x) = \tau(x, x + \Delta x)[I - R(0, x)\mathscr{R}(x, x + \Delta x)]^{-1}\tau(0, x), \tag{3.17a}$$

$$T(0, x + \Delta x) = T(0, x)[I - \mathscr{R}(x, x + \Delta x)R(0, x)]^{-1}T(x, x + \Delta x), \tag{3.17b}$$

$$\mathscr{R}(0, x + \Delta x) = T(0, x)[I - \mathscr{R}(x, x + \Delta x)R(0, x)]^{-1}\mathscr{R}(x, x + \Delta x) \\ \cdot \tau(0, x) + \mathscr{R}(0, x), \tag{3.17c}$$

$$R(0, x + \Delta x) = \tau(x, x + \Delta x)[I - R(0, x)\mathscr{R}(x, x + \Delta x)]^{-1}R(0, x) \\ \cdot T(x, x + \Delta x) + R(x, x + \Delta x). \tag{3.17d}$$

Provided that the four incremental coefficients are known, the recursive equations can be used to find the composite coefficients for the line. The equations are used in the same manner as the discrete case equations.

The initial conditions on R, T, $\mathscr{R}$, and τ are the the same as given previously. In addition to the initial conditions for $x = 0$,

$$R(0, 0) = \mathscr{R}(0, 0) = 0,$$

$$T(0, 0) = \tau(0, 0) = I,$$

a second set of initial conditions are necessary. If $R(x, x)$, $\mathscr{R}(x, x)$, $T(x, x)$, and $\tau(x, x)$ indicate the reflection and transmission coefficients for a zero thickness slab at x, it is not difficult to see that

$$R(x, x) = \mathscr{R}(x, x) = 0,$$

$$T(x, x) = \tau(x, x) = I.$$

The equations above merely imply that a slab of zero thickness must have a transmissivity of unity.

4. MEDIUM THICKNESS INCREASING IN THE DIRECTION OF NEGATIVE x

There may be times when the medium thickness must be increased in the negative x direction. The signal flow graph can easily be modified to handle this case and the recursive equations derived. Consider the graph shown in Figure 3.9, where the slab of thickness Δx has to be adjoined to the left side

3 RECURSIVE EQUATIONS

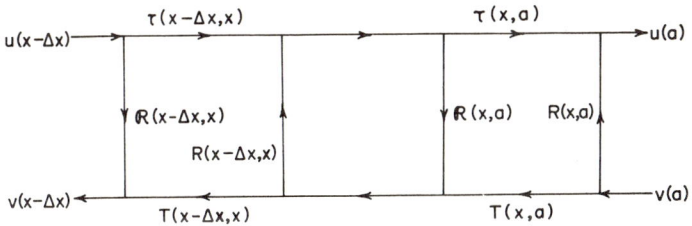

Figure 3.9. Signal flow graph for negative x.

of the medium. The recursive equations can be found by inspection.

$$\tau(-x - \Delta x, 0) = \tau(-x, 0)[I - R(-x - \Delta x, -x)\mathscr{R}(-x, 0)]^{-1} \\ \cdot \tau(-x - \Delta x, -x), \tag{3.18a}$$

$$T(-x - \Delta x, 0) = T(-x - \Delta x, -x)[I - \mathscr{R}(-x, 0)R(-x - \Delta x, -x)]^{-1} \\ \cdot T(-x, 0), \tag{3.18b}$$

$$R(-x - \Delta x, 0) = \tau(-x, 0)[I - R(-x - \Delta x, -x)\mathscr{R}(-x, 0)]^{-1} \\ \cdot R(-x - \Delta x, -x)T(-x, 0) + R(-x, 0), \tag{3.18c}$$

$$\mathscr{R}(-x - \Delta x, 0) = T(-x - \Delta x, -x)[I - \mathscr{R}(-x, 0)R(-x - \Delta x, -x)]^{-1} \\ \cdot \mathscr{R}(-x, 0)\tau(-x - \Delta x, -x) + \mathscr{R}(-x - \Delta x, -x). \tag{3.18d}$$

It is obvious from the flow graph how the equations are modified for the discrete case, which was discussed earlier. The reader should also consider the case where the medium is increasing in thickness in the direction of decreasing x but with x entirely positive.

5. CHANGE IN THE DIRECTION OF FLOW

The recursive equations which have been derived were based on the flow of $u(x)$ to the right and $v(x)$ to the left. The recursive equations can be derived for the reversal of the direction of the two waves. The signal flow graph is again used since the equations can be obtained by inspection. Consider the flow graph given below where $u(x)$ is the flow to the left and $v(x)$ the flow to the right. The medium thickness is increasing in the direction of positive x and the medium is assumed to be continuous.

The four recursive equations can be obtained directly from Figure 3.10 by following the same rules as used earlier. The tilde ($\sim$) above the

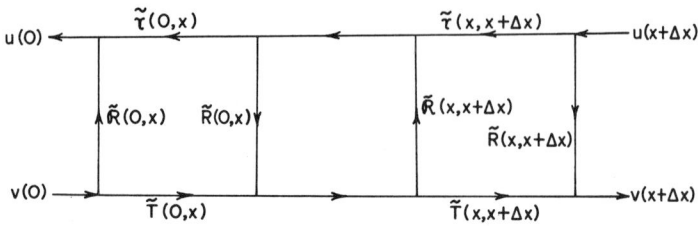

Figure 3.10. Signal flow graph for opposite flow.

coefficients indicates the coefficients for the change of direction of flow.

$$\tilde{\tau}(0, x + \Delta x) = \tilde{\tau}(0, x)[I - \tilde{\mathscr{R}}(x, x + \Delta x)\tilde{R}(0, x)]^{-1}\tilde{\tau}(x, x + \Delta x), \tag{3.19a}$$

$$\tilde{T}(0, x + \Delta x) = \tilde{T}(x, x + \Delta x)[I - \tilde{R}(0, x)\tilde{\mathscr{R}}(x, x + \Delta x)]^{-1}\tilde{T}(0, x), \tag{3.19b}$$

$$\tilde{R}(0, x + \Delta x) = \tilde{T}(x, x + \Delta x)[I - \tilde{R}(0, x)\tilde{\mathscr{R}}(x, x + \Delta x)]^{-1}$$
$$\cdot \tilde{R}(0, x)\tilde{\tau}(x, x + \Delta x) + \tilde{R}(x, x + \Delta x), \tag{3.19c}$$

$$\tilde{\mathscr{R}}(0, x + \Delta x) = \tilde{\tau}(0, x)[I - \tilde{\mathscr{R}}(x, x + \Delta x)\tilde{R}(0, x)]^{-1}$$
$$\cdot \tilde{\mathscr{R}}(x, x + \Delta x)\tilde{T}(0, x) + \tilde{\mathscr{R}}(0, x). \tag{3.19d}$$

The recursive equations are modified from the ones given earlier. If (3.19) is compared to (3.17), note that the equations are of identical form if these substitutions are made:

$$\tilde{\tau}(0, x + \Delta x) = T(0, x + \Delta x), \tag{3.20a}$$
$$\tilde{T}(0, x + \Delta x) = \tau(0, x + \Delta x), \tag{3.20b}$$

and

$$\tilde{\mathscr{R}}(0, x + \Delta x) = \mathscr{R}(0, x + \Delta x), \tag{3.20c}$$
$$\tilde{R}(0, x + \Delta x) = R(0, x + \Delta x), \tag{3.20d}$$

in the recursive equations of (3.19). This does not imply that the coefficients obtained from (3.19) are identical with those of (3.17), and the definitions of (3.20) hold. It is shown later that the differential equations for the two reflection and the two transmission coefficients are not the same for the two flows.

The equations for $u(x)$ and $v(x)$ are also modified from the ones given in (3.15):

$$u(0) = \tilde{\tau}(0, x)u(x) + \tilde{\mathscr{R}}(0, x)v(0), \tag{3.21a}$$
$$v(x) = \tilde{R}(0, x)u(x) = \tilde{T}(0, x)v(0). \tag{3.21b}$$

3 RECURSIVE EQUATIONS

It can be shown from (3.21) and (3.15) that the positive $u(x)$ flow coefficients and the positive $v(x)$ coefficients are interrelated as given below.

$$\tilde{\tau}(0, x) = [I - \tilde{\tau}^{-1}(0, x)\tilde{\mathscr{R}}(0, x)\tilde{T}^{-1}(0, x)\tilde{R}(0, x)]^{-1}\tilde{\tau}^{-1}(0, x), \quad (3.22a)$$

$$\tilde{R}(0, x) = -[I - \tilde{\tau}^{-1}(0, x)\tilde{\mathscr{R}}(0, x)\tilde{T}^{-1}(0, x)\tilde{R}(0, x)]^{-1}\tilde{\tau}^{-1}(0, x)$$
$$\cdot \tilde{\mathscr{R}}(0, x)\tilde{T}^{-1}(0, x), \quad (3.22b)$$

$$\mathscr{R}(0, x) = \tilde{T}^{-1}(0, x)\tilde{R}(0, x)[I - \tilde{\tau}^{-1}(0, x)\tilde{\mathscr{R}}(0, x)\tilde{T}^{-1}(0, x)\tilde{R}(0, x)]^{-1}$$
$$\cdot \tilde{\tau}^{-1}(0, x), \quad (3.22c)$$

$$T(0, x) = -\tilde{T}^{-1}(0, x)\tilde{R}(0, x)[I - \tilde{\tau}^{-1}(0, x)\tilde{\mathscr{R}}(0, x)\tilde{T}^{-1}(0, x)\tilde{R}(0, x)]^{-1}$$
$$\cdot \tilde{\tau}^{-1}(0, x)\tilde{\mathscr{R}}(0, x)\tilde{T}^{-1}(0, x) - \tilde{T}^{-1}(0, x). \quad (3.22d)$$

It can be seen from (3.22) that the tilde coefficients are quite different from the coefficients for the positive $u(x)$ flow.

Recursive equations for a medium of thickness increasing in the direction of negative x can be obtained by making suitable changes in the signal flow graph of Figure 3.10.

6. THE MATRIX PRODUCT FORMULATION

The recursive equations of (3.14), (3.17), (3.18), and (3.19) are useful in computing the reflection, transmission, and coupling coefficients for a medium. There are certain problems which may give rise to computational difficulties when the transmission matrix is calculated. In the one-species neutron problem, the reflection and transmission coefficients approach an infinite value at the critical length. An alternative method of finding the values of $u(x)$ and $v(x)$ can be derived which avoids the singular points of the transmission matrix. The matrix product formulation is one method of avoiding the infinite value of the reflection and transmission coefficients, although this method has undesirable features.

To distinguish between the recursive equations of the previous sections, the equations developed earlier are now called the transmission matrix recursive equations. The equations derived in this section are referred to as the fundamental matrix product equations. Brillouin [8] has discussed the differential equations for the matrix product, and the reader will find this discussion useful in analyzing problems by this method. The differential equations for the transmission matrix and the fundamental matrix are discussed in the next chapter.

One of the difficulties in using the transmission matrix recursive equations is the presence of the matrix sum $[I - \mathscr{R}(x, x + \Delta x)R(0, x)]$ in the equations. Since the inverse of this matrix must be found, it must always be nonsingular. The nonsingular requirement cannot always be assured since $R(0, x)$ may approach infinity. This requires that the transmission matrix method be restricted to problems where

$$\mathscr{R}(x, x + \Delta x)R(0, x) \neq I.$$

The fundamental matrix method, as will be shown, requires that $T(0, x)$ be nonsingular.

The transmission matrix equations can be modified in a straightforward manner to obtain the fundamental matrix equations. The transmission matrix equation for a length of line, x, is given by

$$\begin{bmatrix} u(x) \\ v(0) \end{bmatrix} = \begin{bmatrix} \tau(0, x) & R(0, x) \\ \mathscr{R}(0, x) & T(0, x) \end{bmatrix} \begin{bmatrix} u(0) \\ v(x) \end{bmatrix}. \quad (3.23)$$

The fundamental matrix equation is

$$\begin{bmatrix} u(x) \\ v(x) \end{bmatrix} = \begin{bmatrix} Q_{11}(0, x) & Q_{12}(0, x) \\ Q_{21}(0, x) & Q_{22}(0, x) \end{bmatrix} \begin{bmatrix} u(0) \\ v(0) \end{bmatrix}, \quad (3.24)$$

where $u(0)$ and $v(0)$ are now independent parameters. To obtain $\mathbf{Q}(0, x)$ from (3.23), the equations must be manipulated into the form of (3.24). This can easily be carried out, and

$$\mathbf{Q}(0, x) = \begin{bmatrix} \tau(0, x) - R(0, x)T^{-1}(0, x)\mathscr{R}(0, x) & R(0, x)T^{-1}(0, x) \\ -T^{-1}(0, x)\mathscr{R}(0, x) & T^{-1}(0, x) \end{bmatrix}. \quad (3.25)$$

Thus, if $\mathbf{T}(0, x)$ (the transmission matrix) is known, $\mathbf{Q}(0, x)$ can easily be found. Similarly, the incremental fundamental matrix $\mathbf{Q}(x, x + \Delta x)$ is found by making use of (3.25).

$$\mathbf{Q}(x, x + \Delta x)$$
$$= \begin{bmatrix} \tau(x, x + \Delta x) - R(x, x + \Delta x) & R(x, x + \Delta x)T^{-1}(x, x + \Delta x) \\ \times T^{-1}(x, x + \Delta x)\mathscr{R}(x, x + \Delta x) & \\ -T^{-1}(x, x + \Delta x)\mathscr{R}(x, x + \Delta x) & T^{-1}(x, x + \Delta x) \end{bmatrix}. \quad (3.26)$$

Using the definition of the fundamental matrix for an incremental distance Δx,

$$\begin{bmatrix} u(x + \Delta x) \\ v(x + \Delta x) \end{bmatrix} = \mathbf{Q}(x, x + \Delta x) \begin{bmatrix} u(x) \\ v(x) \end{bmatrix}, \quad (3.27)$$

3 RECURSIVE EQUATIONS

it can be shown that the functions $u(x + \Delta x)$ and $v(x + \Delta x)$ are given by

$$\begin{bmatrix} u(x + \Delta x) \\ v(x + \Delta x) \end{bmatrix} = \mathbf{Q}(x, x + \Delta x)\mathbf{Q}(0, x)\begin{bmatrix} u(0) \\ v(0) \end{bmatrix}. \quad (3.28)$$

The recursive equations for $\mathbf{Q}(0, x + \Delta x)$ are then found by a matrix multiplication of $\mathbf{Q}(x, x + \Delta x)$ and $\mathbf{Q}(0, x)$.

$$Q_{11}(0, x + \Delta x) = Q_{11}(x, x + \Delta x)Q_{11}(0, x) + Q_{12}(x, x + \Delta x)Q_{21}(0, x), \quad (3.29a)$$

$$Q_{12}(0, x + \Delta x) = Q_{11}(x, x + \Delta x)Q_{12}(0, x) + Q_{12}(x, x + \Delta x)Q_{22}(0, x), \quad (3.29b)$$

$$Q_{21}(0, x + \Delta x) = Q_{21}(x, x + \Delta x)Q_{11}(0, x) + Q_{22}(x, x + \Delta x)Q_{21}(0, x), \quad (3.29c)$$

$$Q_{22}(0, x + \Delta x) = Q_{21}(x, x + \Delta x)Q_{12}(0, x) + Q_{22}(x, x + \Delta x)Q_{22}(0, x). \quad (3.29d)$$

The initial conditions for $\mathbf{Q}(0, 0)$ can be obtained from (3.25) by letting $x \to 0$, and recalling that $\tau(0, 0) = T(0, 0) = I$, $R(0, 0) = \mathcal{R}(0, 0) = 0$.

$$\mathbf{Q}(0, 0) = \begin{bmatrix} I & 0 \\ 0 & I \end{bmatrix}. \quad (3.30)$$

The fundamental matrix $\mathbf{Q}(x, x)$ at any point x is also given by (3.30).

The fundamental matrix can be solved by a recursive method where the fundamental matrix is defined in terms of the transmission matrix. As can be seen from (3.25), the inverse of $T(0, x)$ enters into the equations. If $T(0, x) \to \infty$, as may be the case for the scalar problem, $Q_{22}(0, x)$ is zero. The other terms of (3.25) are complicated by the presence of the other coefficients. The behavior of $T(0, x)$ and $\mathbf{Q}(0, x)$ must be examined in detail before any conclusions can be reached concerning the choice of computational methods.

7. RECURSIVE RELATIONS FOR FORCING FUNCTIONS

The presence of forcing functions in (1.1) and (1.2) has been neglected in deriving the recursive equations. Recursive equations for the forcing function's contribution to the solutions $u(x)$ and $v(x)$ can be generated. A simple modification of the signal flow graph is required to account for the forcing functions, and the derivation of the recursive equations is now given.

Assume that a new "source" term is added to (3.15a) and (3.15b) which is the contribution of the forcing functions to the functions $u(x)$ and $v(x)$. The differential equations for $u(x)$ and $v(x)$ are

$$\frac{du(x)}{dx} = B(x)u(x) + A(x)v(x) + e(x), \tag{1.1}$$

$$-\frac{dv(x)}{dx} = D(x)u(x) + C(x)v(x) + f(x). \tag{1.2}$$

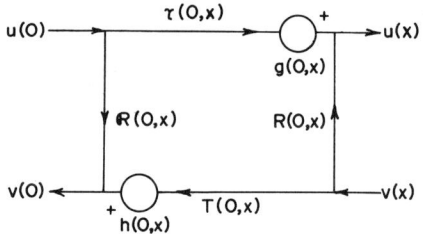

Figure 3.11. Signal flow graph with sources.

the "field" on each side of the medium as defined below.

$$u(x) = \tau(0, x)u(0) + R(0, x)v(x) + g(0, x), \tag{3.31a}$$
$$v(0) = \mathscr{R}(0, x)u(0) + T(0, x)v(x) + h(0, x). \tag{3.31b}$$

The signal flow graph for (3.31a) and (3.31b) must now include a source generator; the graph is given in Figure 3.11. The source generators are assumed to be "short-circuited" to the "flow" through the graph and are placed within the medium. An "open-circuited" generator could also be placed in parallel to $\tau(0, x)$ and $T(0, x)$.

A cascaded flow graph can now be drawn with a medium of thickness Δx placed to the right of the medium extending from 0 to x.

The recursive equations for use when forcing functions are present follow directly from Figure 3.12. The "flow" $u(x + \Delta x)$ consists of the contributions from $u(0)$ and $v(x + \Delta x)$ as well as that of the internal generators.

$$\begin{aligned}
u(x + \Delta x) &= \tau(x, x + \Delta x)[I - R(0, x)\mathscr{R}(x, x + \Delta x)]^{-1}\tau(0, x)u(0) \\
&+ \{\tau(x, x + \Delta x)[I - R(0, x)\mathscr{R}(x, x + \Delta x)]^{-1} \\
&\times R(0, x)T(x, x + \Delta x) + R(x, x + \Delta x)\}v(x + \Delta x) \\
&+ \{g(x, x + \Delta x) + \tau(x, x + \Delta x) \\
&\times [I - R(0, x)\mathscr{R}(x, x + \Delta x)]^{-1}g(0, x) + \tau(x, x + \Delta x) \\
&\times [I - R(0, x)\mathscr{R}(x, x + \Delta x)]^{-1}R(0, x)h(x, x + \Delta x)\}.
\end{aligned} \tag{3.32a}$$

3 RECURSIVE EQUATIONS

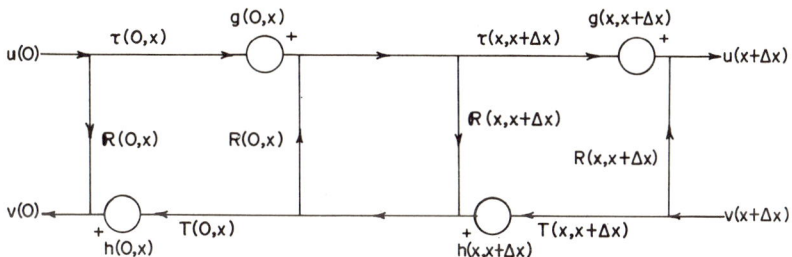

Figure 3.12. Cascade signal flow graph with sources.

The function $v(0)$ can be written down directly from the graph:

$$\begin{aligned}
v(0) = &\{T(0, x)[I - \mathscr{R}(x, x + \Delta x)R(0, x)]^{-1} \\
&\times \mathscr{R}(x, x + \Delta x)\tau(0, x) + \mathscr{R}(0, x)\}u(0) \\
&+ \{T(0, x)[I - \mathscr{R}(x, x + \Delta x)R(0, x)]^{-1}T(x, x + \Delta x)\}v(x + \Delta x) \\
&+ \{h(0, x) + T(0, x)[I - \mathscr{R}(x, x + \Delta x)R(0, x)]^{-1}h(x, x + \Delta x) \\
&+ T(0, x)[I - \mathscr{R}(x, x + \Delta x)R(0, x)]^{-1}\mathscr{R}(x, x + \Delta x)g(0, x)\}.
\end{aligned}$$
(3.32b)

Equations (3.31a) and (3.31b) are valid for a medium of thickness x, if the thickness is increased to $x + \Delta x$, (3.31) still holds but with x replaced by $x + \Delta x$. It can be seen immediately that the first two terms in (3.32a) are $\tau(0, x + \Delta x)$ and $R(0, x + \Delta x)$, respectively. These composite coefficients are unchanged. Similarly, (3.32b) has $\mathscr{R}(0, x + \Delta x)$ and $T(0, x + \Delta x)$ as the leading two terms on the right side. The third term on the right side of (3.32a) is $g(0, x + \Delta x)$, and the final term of (3.32b) is $h(0, x + \Delta x)$.

$$\begin{aligned}
g(0, x + \Delta x) = &g(x, x + \Delta x) + \tau(x, x + \Delta x)[I - R(0, x)\mathscr{R}(x, x + \Delta x)]^{-1} \\
&\times g(0, x) + \tau(x, x + \Delta x)[I - R(0, x)\mathscr{R}(x, x + \Delta x)]^{-1} \\
&\times R(0, x)h(x, x + \Delta x),
\end{aligned}$$
(3.33a)

$$\begin{aligned}
h(0, x + \Delta x) = &h(0, x) + T(0, x)[I - \mathscr{R}(x, x + \Delta x)R(0, x)]^{-1}h(x, x + \Delta x) \\
&+ T(0, x)[I - \mathscr{R}(x, x + \Delta x)R(0, x)]^{-1} \\
&\times \mathscr{R}(x, x + \Delta x)g(0, x).
\end{aligned}$$
(3.33b)

The source functions defined in (3.33) are valid when the forcing functions $e(x)$ and $f(x)$ are present. As would be expected, the functions depend on the transmission matrix elements as well as on the source incremental coefficients, $g(x, x + \Delta x)$ and $h(x, x + \Delta x)$. These new source incremental coefficients depend on the forcing functions $e(x)$ and $f(x)$ as well as on the matrices $A(x)$, $B(x)$, $C(x)$, and $D(x)$.

The differential equations for $g(0, x)$ and $h(0, x)$ are derived in the next chapter and the incremental coefficients are discussed.

8. SUMMARY

Recursive equations have been derived in this chapter for cascading sections of a medium together to find the composite transmission matrix. The sections of "line" may be discrete with constant "impedance," or the line may have an impedance which varies along each section. The forms of the recursive equations are the same, the physical meaning of the functions which appear in the equations differs in the discrete and continuous cases. The concept of incremental coefficients has been introduced.

It was shown that internal sources are included by considering two additional recursive functions. The recursive equations for the transmission matrix is unchanged when the sources are added. The source recursive equations account for creation of waves in the medium by some type of excitation.

REFERENCES

1. R. N. Adams and E. Denman, *Wave Propagation in Turbulent Media*, American Elsevier Co., New York, 1967.
2. R. Bellman, R. Kalaba, and G. M. Wing, *Invariant Imbedding and Mathematical Physics I; Particle Processes*. *J. Math. Phys.*, *1* (1960), 280.
3. R. Bellman and R. Kalaba, Wave Branching Processes and Invariant Imbedding, *Proc. Natl. Acad. Sci. U.S.*, *47* (1961), 1507.
4. A. Shimizu and H. Mizuta, Application of Invariant Imbedding to the Reflection and Transmission Problem of Gamma Ray, I, *J. Nucl. Sci. Technol. (Japan)*, *3*, 1966.
5. R. Redheffer, On the Relation of Transmission Line Theory to Scattering and Transfer, *J. Math. Phys.*, *41* (1962), 1.
6. W. T. Reid, Solution of a Riccati Matrix Differential Equation as Functions of Initial Values, *J. Math. Mech.*, *8* (1959), 221.
7. R. Adams and E. Denman, A Precautionary Note on Stratification, *Radio Sci.*, *1* (1966), 851.
8. L. Brillouin, *Wave Propagation in Periodic Structures*, Dover Publication, New York, 1953.

RELATED REFERENCES

R. Bellman, Functional Equations, Wave Propagation and Invariant Imbedding, *J. Math. Mech.*, *8* (1959), 683.

R. Bellman and R. Kalaba, Invariant Imbedding and Wave Propagation in Stochastic Media, *Electromagnetic Wave Propagation*, Academic Press, New York, 1960.

H. Davies and E. Denman, Invariant Imbedding as a Basic for the Solution of Quantum Mechanical Problems, *J. Math. Anal. and Appl.*, *19* (1967), 133.

E. Denman and J. P. Kelley, Invariant Imbedding and the Band Theory of Solids, *J. Math. Anal. and Appl.*, *21* (1968).

E. Denman, Coupled Mode Theory, *J. Math. Anal. and Appl.*, *21* (1968).

R. W. Preisendorfer, *Radiative Transfer on Discrete Spaces*, Pergamon Press, New York, 1965.

CHAPTER 4

The Riccati, Recursive, and Other Differential Equations

The Riccati differential equation is a well-known mathematical equation which is frequently encountered in mathematical physics. The equation is a nonlinear first-order differential equation which does not have a general solution. The Riccati equation discussed in this chapter is the matrix equation of the form

$$\frac{dR(x)}{dx} = A(x) + B(x)R(x) + R(x)C(x) + R(x)D(x)R(x), \qquad (4.1)$$

where $A(x)$, $B(x)$, $C(x)$, and $D(x)$ are $n \times n$ matrices; $R(x)$ is an n-order vector. Many linear differential equations can be transformed in a Riccati equation, although this is seldom done. The reason is readily apparent: it is a difficult equation with which to work since it does not have a closed-form solution in the most general case.

For example, the second-order differential equation

$$y''(x) + p(x)y'(x) + Q(x)y(x) = 0 \qquad (4.2)$$

leads to the Riccati equation when a simple change of variable is made. Let

$$R(x) = \frac{y(x)}{y'(x)},$$

then

$$\frac{dR}{dx} = 1 + p(x)R(x) + R(x) + R(x)q(x)R(x). \qquad (4.3)$$

Equation (4.3) is more difficult to solve than is (4.2); thus the second-order equation is preferred. Second-order equations of the type of (4.2) are encountered in wave propagation work. If a means of solving the Riccati equation is available, then the second-order equation can be solved by using $R(x)$.

4 THE RICCATI, RECURSIVE, AND OTHER DIFFERENTIAL EQUATIONS

The Riccati differential equation can be solved numerically in several ways (see Bellman and Kalaba [1] or any book on numerical analysis). No attempt is made here to solve the Riccati equation analytically or to improve on the numerical methods in current usage. The emphasis is placed on showing that the recursive equations lead to the Riccati equation and on devising a numerical scheme based on the recursive equations. The four differential equations for the reflection and transmission coefficients have been studied by Reid [2] and Redheffer [3]. The solution of the four differential equations for the transmission matrix coefficients determines the propagation characteristic of the medium. If the coefficients can be computed everywhere, $u(x)$ and $v(x)$ can be obtained for the coupled differential equations.

The goal set for this chapter is to obtain the Riccati equation for the reflection coefficient from the matrix recursive equations and the three other differential equations. The mathematical relationships between first- and second-order differential equations, the coupled differential equations, the Riccati equation, and the recursive equations are presented. The recursive relations are useful in solving coupled mode problems as well as ordinary differential equations. Numerous differential equations solved numerically are presented in a later chapter.

I. HOMOGENEOUS SECOND-ORDER DIFFERENTIAL EQUATIONS AND THE COUPLED FIRST-ORDER DIFFERENTIAL EQUATIONS

As a starting point for the analysis, the second-order differential equation

$$\frac{d^2 y(x)}{dx^2} + p(x)\frac{dy(x)}{dx} + q(x)y(x) = 0 \tag{4.2}$$

is considered. This equation was discussed in the introduction to this chapter and is of interest in wave propagation, circuit analysis, and other engineering problems. The differential equation describes the state of an oscillatory system where $p(x)$ is the damping coefficient and $q(x)$ contains the frequency information of the system.

Equation (4.2) can be reduced to a set of first-order coupled differential equations such as given in (1.1) and (1.2) by making simple changes. Let

$$y(x) = u(x), \tag{4.3}$$

$$\frac{dy(x)}{dx} = v(x), \tag{4.4}$$

and substitute back in (4.2). Equation (4.4) describes one of the equations, and the other is given by (4.2) with the substitutions from (4.3) and (4.4).

$$\frac{du(x)}{dx} = v(x), \qquad (4.5a)$$

$$-\frac{dv(x)}{dx} = -q(x)u(x) + p(x)v(x). \qquad (4.5b)$$

The second-order differential equation of (4.2) can be stated in terms of a coupled set, where the coupling takes place between the function $y(x)$ and its derivative $y'(x)$.

Higher-order differential equations can be treated in the same manner as the second-order equation above. The coupled equations are of even order if $A(x)$, $B(x)$, $C(x)$, and $D(x)$ are square matrices of order $r \times r$. If the set of coupled equations is odd, an uncoupled arbitrary equation can be adjoined to the set to give an even set. It is also possible to increase the order of the equations by considering a null solution.

2. THE COUPLED FIRST-ORDER EQUATIONS AND THE RICCATI EQUATION

It was shown in the last section that a simple second-order differential equation can be expressed in terms of a coupled differential equation set. The coefficients of the coupled set can be identified from the original second-order equation. The Riccati equation can then be derived from the coupled set. Suppose that $v(x)$ and $u(x)$ are amplitude functions, where $v(x)$ defines the "flow" to the left and $u(x)$ the "flow" to the right. The functions $u(0)$ and $v(x)$ are the incident amplitudes on the left and the right sides of the medium. Let $u(x)$ and $v(x)$ be related to each other through the equation

$$u(x) = \tau(0, x)u(0) + R(0, x)v(x). \qquad (4.6)$$

The coupled set of differential equations is given by

$$\frac{du(x)}{dx} = B(x)u(x) + A(x)v(x), \qquad (3.1a)$$

$$-\frac{dv(x)}{dx} = D(x)u(x) + C(x)v(x), \qquad (3.1b)$$

where it is assumed that $A(x)$, $B(x)$, $C(x)$, and $D(x)$ are bounded. The negative sign of $dv(x)/dx$ is important, and care must be taken to assure that

4 THE RICCATI, RECURSIVE, AND OTHER DIFFERENTIAL EQUATIONS

(3.1) is always defined with the proper signs. Now differentiate (4.6) with respect to x:

$$\frac{du(x)}{dx} = \frac{d\tau(0, x)}{dx} u(0) + \frac{dR(0, x)}{dx} v(x) + R(0, x) \frac{dv(x)}{dx}. \quad (4.7)$$

The function $u(0)$ has been considered as a constant since it is at the left-hand boundary of the medium and is the input on that side.

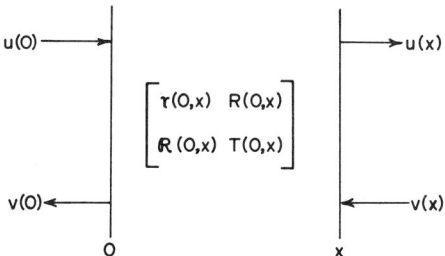

Figure 4.1. Boundary condition model.

Substitute from (3.1a) and (3.1b) for $du(x)/dx$ and $dv(x)/dx$. Carrying out the substitution and rearranging, the following equation holds.

$$\left[\frac{dR(0, x)}{dx} - A(x) - B(x)R(0, x) - R(0, x)C(x) - R(0, x)D(x)R(0, x)\right]v(x)$$

$$= \left[B(x)\tau(0, x) + R(0, x)D(x)\tau(0, x) - \frac{d\tau(0, x)}{dx}\right]u(0). \quad (4.8)$$

To interpret (4.8), consider the wave picture in Figure 4.1. Functions $u(0)$ and $v(x)$ are inputs to the system; and, since $u(0)$ is a constant, $u(0)$ and $v(x)$ can be treated as independent variables. Equation (4.8) can thus be viewed as two separate equations. If $v(x) = 0$, the right side of (4.8) must hold; that is,

$$\frac{d\tau(0, x)}{dx} = [B(x) + R(0, x)D(x)]\tau(0, x). \quad (4.9a)$$

On the other hand, if $u(0)$ is zero, the left side of (4.8) still holds:

$$\frac{dR(0, x)}{dx} = A(x) + B(x)R(0, x) + R(0, x)C(x) + R(0, x)D(x)R(0, x).$$

$$(4.9b)$$

Although the derivation above lacks mathematical rigor, the derivation based on the argument is valid. A rigorous derivation of the two equations of (4.9) can be found in Bailey and Wing [4].

The existence of a "reflection" and "transmission" differential equation for (4.6) suggests that an analogous set should exist for the remaining two coefficients of the transmission matrix. To find the two differential equations for the other two coefficients, differentiate the equation

$$v(0) = \mathscr{R}(0, x)u(0) + T(0, x)v(x) \tag{4.10}$$

and substitute into the resulting equation to eliminate $du(x)/dx$ and $v(x)/dx$. When this is done,

$$\left[\frac{d\mathscr{R}(0, x)}{dx} - T(0, x)D(x)\tau(0, x)\right]u(0)$$
$$+ \left[\frac{dT(0, x)}{dx} - T(0, x)D(x)R(0, x) - T(0, x)C(x)\right]v(x) = 0. \tag{4.11}$$

The same argument as given prior to (4.9a) holds. The function $u(0)$ can be zero independently of $v(x)$; therefore,

$$\frac{dT(0, x)}{dx} = T(0, x)[D(x)R(0, x) + C(x)], \tag{4.9c}$$

$$\frac{d\mathscr{R}(0, x)}{dx} = T(0, x)D(x)\tau(0, x). \tag{4.9d}$$

The equations given in (4.9) are the set of four differential equations for the four coefficients of the transmission matrix of (3.3).

The four differential equations of (4.9) form a complete set analogous to the four recursive equations given earlier. Some general comments should be made regarding the four differential equations of (4.9) before showing that the recursive equations generate differential equations identical with those of (4.9) in the limits as $\Delta x \to 0$. First, note that the Riccati equation for $R(0, x)$ is completely independent of the three differential equations for $\tau(0, x)$, $T(0, x)$, and $\mathscr{R}(0, x)$. The initial conditions $R(x, x) = \mathscr{R}(x, x) = 0$ and $T(x, x) = \tau(x, x) = I$ still hold, and (4.9a) can be solved without considering the other three equations. The differential equations for $T(0, x)$ and $\tau(0, x)$ do not depend on $\mathscr{R}(0, x)$; thus they can be solved if $R(0, x)$ is known. The equation for $\mathscr{R}(0, x)$ depends on the three parameters $R(0, x)$, $T(0, x)$, and $T(0, x)$; thus it cannot be solved until these coefficients are found.

If $u(0) = 0$, then only two of the differential equations of (4.9) must be solved to find $u(x)$ and $v(x)$. For the case where $u(0) = 0$, $u(x)$ is given by

$$u(x) = R(0, x)v(x)$$

and $v(x)$ can be found from (4.10).

$$v(0) = T(0, x)v(x).$$

The coefficients $\mathscr{R}(0, x)$ and $\tau(0, x)$ do not enter into the equations for $v(x)$ and $u(x)$ and are, therefore, of no importance.

3. THE RECURSIVE EQUATIONS AND THE RICCATI EQUATION

The recursive equations derived in Chapter 3 are useful in numerical analysis, as they can be used to find $R(0, x + \Delta x)$, $T(0, x + \Delta x)$, $\tau(0, x + \Delta x)$, and $R(0, x + \Delta x)$ without solving the differential equations for the transmission matrix. The validity of the recursive equations for numerical analysis is now established by deriving the four differential equations of (4.9) directly from the recursive relations. To derive the four differential equations of (4.9), the recursive relations are considered in the limits as Δx approaches zero.

Consider the recursive equation for $R(0, x + \Delta x)$ first.

$$R(0, x + \Delta x) = \tau(x, x + \Delta x)[I - R(0, x)\mathscr{R}(x, x + \Delta x)]^{-1}$$
$$\times R(0, x)T(x, x + \Delta x) + R(x, x + \Delta x). \quad (3.7d)$$

If the medium is increased in thickness by Δx, then $R(x, x + \Delta x)$, $T(x, x + \Delta x)$, $\tau(x, x + \Delta x)$, and $\mathscr{R}(x, x + \Delta x)$ must account for the change in $R(0, x + \Delta x)$. To find the incremental coefficients, consider the coupled differential equations of (2.1).

$$\frac{du(x)}{dx} = B(x)u(x) + A(x)v(x), \quad (2.1a)$$

$$-\frac{dv(x)}{dx} = D(x)u(x) + C(x)v(x). \quad (2.1b)$$

Now, from the Taylor series expansion of $u(x + \Delta x)$ and $v(x + \Delta x)$,

$$u(x + \Delta x) = u(x) + B(x)\,\Delta x\, u(x) + A(x)\,\Delta x\, v(x) + \cdots, \quad (4.12)$$
$$v(x + \Delta x) = v(x) - D(x)\,\Delta x\, u(x) - C'(x)\,\Delta x\, v(x) - \cdots. \quad (4.13)$$

Retaining only first-order terms in Δx, (4.12) and (4.13) can be rewritten as

$$u(x + \Delta x) = [I + B(x)\,\Delta x]u(x) + [A(x)\,\Delta x]v(x + \Delta x), \quad (4.14a)$$
$$v(x) = [D(x)\,\Delta x]u(x) + [I + C(x)\,\Delta x]v(x + \Delta x); \quad (4.14b)$$

thus, on comparing (4.14a) to (4.6) and (4.14b) to (4.10), the incremental coefficients for the first-order approximation are

$$R(x, x + \Delta x) = A(x)\,\Delta x, \quad (4.15a)$$
$$T(x, x + \Delta x) = I + C(x)\,\Delta x, \quad (4.15b)$$
$$\tau(x, x + \Delta x) = I + B(x)\,\Delta x, \quad (4.15c)$$
$$\mathcal{R}(x, x + \Delta x) = D(x)\,\Delta x. \quad (4.15d)$$

If Δx passes to zero, using the initial conditions, we find

$$\lim_{\Delta x \to 0} \frac{R(x, x + \Delta x) - R(x, x)}{\Delta x} = A(x), \quad (4.16a)$$

$$\lim_{\Delta x \to 0} \frac{\tau(x, x + \Delta x) - (x, x)}{\Delta x} = B(x), \quad (4.16b)$$

$$\lim_{\Delta x \to 0} \frac{T(x, x + \Delta x) - T(x, x)}{\Delta x} = C(x), \quad (4.16c)$$

$$\lim_{\Delta x \to 0} \frac{\mathcal{R}(x, x + \Delta x) - \mathcal{R}(x, x)}{\Delta x} = D(x). \quad (4.16d)$$

The recursive equations of $R(0, x + \Delta x)$ can now be studied since the incremental coefficients for first-order are now known. To find the differential equation for $R(0, x)$, subtract $R(0, x)$ from both sides of (3.7d) and replace the incremental coefficients by (4.15).

$$R(0, x + \Delta x) - R(0, x) = [I + B(x)\,\Delta x][I - R(0, x)D(x)\,\Delta x]^{-1}R(0, x)$$
$$\cdot [I + C(x)\,\Delta x] + A(x)\,\Delta x - R(0, x). \quad (4.17)$$

Retaining only first-order terms, (4.17) can be written as

$$R(0, x + \Delta x) - R(0, x) = A(x)\,\Delta x + B(x)R(0, x)\,\Delta x + R(0, x)C(x)\,\Delta x$$
$$+ R(0, x)D(x)R(0, x)\,\Delta x. \quad (4.18)$$

Since the mathematical definition of the derivative is

$$\frac{dR(0, x)}{dx} = \lim_{\Delta x \to 0} \frac{R(0, x + \Delta x) - R(0, x)}{\Delta x},$$

4 THE RICCATI, RECURSIVE, AND OTHER DIFFERENTIAL EQUATIONS 53

then (4.18) in the limit gives

$$\frac{dR(0, x)}{dx} = A(x) + B(x)R(0, x) + R(0, x)C(x) + R(0, x)D(x)R(0, x), \quad (4.19)$$

which is the desired differential equation.

The recursive equation for $\tau(0, x + \Delta x)$ can be used to find the differential equation for $\tau(0, x)$; $\tau(0, x + \Delta x)$ is given by (3.17a) and, on using the incremental coefficients,

$$\tau(0, x + \Delta x) - \tau(0, x) = [I + B(x)\Delta x][I - R(0, x)D(x)\Delta x]^{-1}$$
$$\cdot \tau(0, x) - \tau(0, x). \quad (4.20)$$

Keeping only first-order terms, dividing by Δx, and taking the limit, the differential equation for $\tau(0, x)$ is found:

$$\frac{d\tau(0, x)}{dx} = [B(x) + R(0, x)D(x)]\tau(0, x), \quad (4.21)$$

which is again the correct equation.

The other two differential equations,

$$\frac{dT(0, x)}{dx} = T(0, x)[C(x) + D(x)R(0, x)] \quad (4.22)$$

and

$$\frac{d\mathscr{R}(0, x)}{dx} = T(0, x)D(x)\tau(0, x), \quad (4.23)$$

can be found by the same mathematical process.

The recursive equations of Chapter 3 are, therefore, the mathematical equivalence of the four differential equations when $\Delta x \neq 0$. Although only first-order corrections are necessary for deriving the differential equations, higher-order corrections for the incremental coefficients are required when Δx is finite. The higher the corrections, the more accurate are the recursive equations. These corrections are discussed later in this chapter.

4. DIFFERENTIAL EQUATIONS OF THE MATRIX PRODUCT

Although the differential equations for the matrix product are not so useful as the differential equations of Section 3, they are derived here for completeness. The recursive equations for the matrix product are given by

(3.29). Consider first the equation for $Q_{11}(0, x + \Delta x)$, given by (3.29a).

$$Q_{11}(0, x + \Delta x) = Q_{11}(x, x + \Delta x)Q_{11}(0, x) + Q_{12}(x, x + \Delta x)Q_{21}(0, x). \tag{3.29a}$$

Define the coefficients $\varepsilon_{ij}(x)$ in the following manner.

$$\varepsilon_{11}(x) = \lim_{\Delta x \to 0} \frac{Q_{11}(x, x + \Delta x) - Q_{11}(x, x)}{\Delta x}, \tag{4.24a}$$

$$\varepsilon_{12}(x) = \lim_{\Delta x \to 0} \frac{Q_{12}(x, x + \Delta x) - Q_{12}(x, x)}{\Delta x}, \tag{4.24b}$$

$$\varepsilon_{21}(x) = \lim_{\Delta x \to 0} \frac{Q_{21}(x, x + \Delta x) - Q_{21}(x, x)}{\Delta x}, \tag{4.24c}$$

$$\varepsilon_{22}(x) = \lim_{\Delta x \to 0} \frac{Q_{22}(x, x + \Delta x) - Q_{22}(x, x)}{\Delta x}. \tag{4.24d}$$

The initial conditions for the fundamental matrix are

$$Q_{11}(x, x) = Q_{22}(x, x) = I, \tag{4.25}$$
$$Q_{12}(x, x) = Q_{21}(x, x) = 0. \tag{4.26}$$

The equations above are the analogous equations to (4.16).

To find $dQ_{11}(0, x)/dx$, subtract $Q_{11}(0, x)$ from both sides of (3.29a) and substitute into the equation for $Q_{11}(x, x + \Delta x)$ and $Q_{12}(x, x + \Delta x)$.

$$Q_{11}(0, x + \Delta x) - Q_{11}(0, x) = [I + \varepsilon_{11}(x) \Delta x]Q_{11}(0, x) - Q_{11}(0, x)$$
$$+ \varepsilon_{12}(x) \Delta x Q_{21}(0, x). \tag{4.27}$$

Dividing (4.27) by Δx and taking the limit, the differential equation for $Q_{11}(0, x)$ is

$$\frac{dQ_{11}(0, x)}{dx} = \varepsilon_{11}(x)Q_{11}(0, x) + \varepsilon_{12}(x)Q_{21}(0, x). \tag{4.28a}$$

The three differential equations for $Q_{12}(0, x)$, $Q_{21}(0, x)$, and $Q_{22}(0, x)$ are found by carrying out the limiting processes on the three recursive equations for these functions.

$$\frac{dQ_{12}(0, x)}{dx} = \varepsilon_{11}(x)Q_{12}(0, x) + \varepsilon_{12}(x)Q_{22}(0, x), \tag{4.28b}$$

$$\frac{dQ_{21}(0, x)}{dx} = \varepsilon_{21}(x)Q_{11}(0, x) + \varepsilon_{22}(x)Q_{21}(0, x), \tag{4.28c}$$

$$\frac{dQ_{22}(0, x)}{dx} = \varepsilon_{21}(x)Q_{12}(0, x) + \varepsilon_{22}(x)Q_{22}(0, x). \tag{4.28d}$$

4 THE RICCATI, RECURSIVE, AND OTHER DIFFERENTIAL EQUATIONS

Note that the four differential equations of (4.28) must be solved simultaneously and are, therefore, more difficult to handle than the differential equations for the transmission matrix. The recursive equations for the matrix product are not difficult to program and are therefore preferred. The solutions of (4.28) are not considered in this book.

It should be pointed out that the coefficients $\varepsilon_{ij}(x)$ are known from the original differential equations. If the solutions $u(x)$ and $v(x)$ are as given in (3.24), then

$$\frac{du(x)}{dx} = \frac{dQ_{11}(0, x)}{dx} u(0) + \frac{dQ_{12}(0, x)}{dx} v(0) = B(x)u(x) + A(x)v(x), \quad (4.29)$$

$$\frac{dv(x)}{dx} = \frac{dQ_{21}(0, x)}{dx} u(0) + \frac{dQ_{22}(0, x)}{dx} v(0) = -D(x)u(x) - C(x)v(x). \quad (4.30)$$

Substituting for $u(x)$ and $v(x)$ from (3.24), we find the four equations

$$\frac{dQ_{11}(0, x)}{dx} = B(x)Q_{11}(0, x) + A(x)Q_{21}(0, x), \qquad Q_{11}(x, x) = 1, \quad (4.31a)$$

$$\frac{dQ_{12}(0, x)}{dx} = B(x)Q_{12}(0, x) + A(x)Q_{22}(0, x), \qquad Q_{12}(x, x) = 0, \quad (4.31b)$$

$$\frac{dQ_{21}(0, x)}{dx} = -D(x)Q_{11}(0, x) - C(x)Q_{21}(0, x), \qquad Q_{21}(x, x) = 0, \quad (4.31c)$$

$$\frac{dQ_{22}(0, x)}{dx} = -D(x)Q_{12}(0, x) - C(x)Q_{22}(0, x), \qquad Q_{22}(x, x) = 1. \quad (4.31d)$$

Comparing (4.31) and (4.28), the coefficients $\varepsilon_{ij}(x)$ are easily found.

$$\varepsilon_{11}(x) = B(x) \quad \varepsilon_{12}(x) = A(x) \quad \varepsilon_{21}(x) = -D(x) \quad \varepsilon_{22}(x) = -C(x). \quad (4.32)$$

5. DIFFERENTIAL EQUATIONS FOR THE TRANSMISSION MATRIX WHEN $u(x)$ IS THE "FLOW" TO THE LEFT

The recursive equations for $v(x)$ taken in the direction of positive x and for $u(x)$ in the negative x direction were given in Chapter 3. In some cases, numerical stability may be poor for $u(x)$ to the right and $v(x)$ to the left. The stability may improve if the flow is taken in the opposite direction from the usual sense, that is, with $v(x)$ in the direction of positive x. The proper choice of flow is not always obvious, and numerical analysis may be necessary to select the proper direction.

The differential equations for the transmission matrix and the incremental coefficients must be known before the reverse flow is used. The original coupled differential equations given at the beginning of the book are again assumed:

$$\frac{du(x)}{dx} = B(x)u(x) + A(x)v(x), \tag{2.1a}$$

$$-\frac{dv(x)}{dx} = D(x)u(x) + C(x)v(x). \tag{2.1b}$$

The coupled differential equations can be written with the negative sign assigned to $du(x)/dx$, but the original form is retained here.

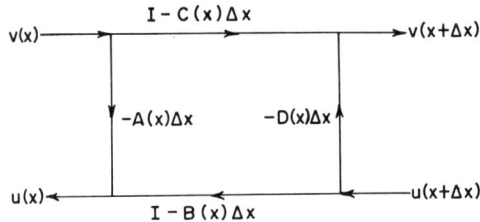

Figure 4.2. Incremental flow for reverse flow.

Assume that the flow of $u(x)$ and that of $v(x)$ are taken as shown in Figure 3.10. Expanding (2.1) in a series and retaining only first-order terms yields

$$u(x) = [I - B(x)\,\Delta x]u(x + \Delta x)A(x)\,\Delta x v(x), \tag{4.33a}$$

$$v(x + \Delta x) = -D(x)\,\Delta x u(x + \Delta x) + [I - C(x)\,\Delta x]v(x), \tag{4.33b}$$

where $u(x + \Delta x)$ and $v(x)$ are the inputs to the system. The signal flow graph for (4.33) is shown in Figure 4.2. The first-order incremental coefficients can now be identified from (4.33) and the equations

$$u(x) = \tilde{\tau}(x, x + \Delta x)u(x + \Delta x) + \tilde{\mathscr{R}}(x, x + \Delta x)v(x), \tag{4.34a}$$

$$v(x + \Delta x) = \tilde{R}(x, x + \Delta x)u(x + \Delta x) + \tilde{T}(x, x + \Delta x)v(x). \tag{4.34b}$$

The first-order incremental coefficients are

$$\tilde{\tau}(x, x + \Delta x) = I - B(x)\,\Delta x, \tag{4.35}$$

$$\tilde{T}(x, x + \Delta x) = I - C(x)\,\Delta x, \tag{4.36}$$

$$\tilde{R}(x, x + \Delta x) = -D(x)\,\Delta x, \tag{4.37}$$

$$\tilde{\mathscr{R}}(x, x + \Delta x) = -A(x)\,\Delta x, \tag{4.38}$$

and, for the limit as $\Delta x \to 0$,

$$\lim_{\Delta x \to 0} \frac{\tilde{\tau}(x, x + \Delta x) - \tilde{\tau}(x, x)}{\Delta x} = -B(x), \quad (4.39a)$$

$$\lim_{\Delta x \to 0} \frac{\tilde{T}(x, x + \Delta x) - \tilde{T}(x, x)}{\Delta x} = -C(x), \quad (4.39b)$$

$$\lim_{\Delta x \to 0} \frac{\tilde{R}(x, x + \Delta x) - \tilde{R}(x, x)}{\Delta x} = -D(x), \quad (4.39c)$$

$$\lim_{\Delta x \to 0} \frac{\tilde{\mathscr{R}}(x, x + \Delta x) - \tilde{\mathscr{R}}(x, x)}{\Delta x} = -A(x). \quad (4.39d)$$

The differential equations for the transmission matrix coefficients with reversed flow can now be derived by following the procedure of Section 3. Substituting the incremental coefficients above in the recursive equations and taking the limit, the differential equations below are derived.

$$\frac{d\tilde{R}(0, x)}{dx} = -D(x) - C(x)\tilde{R}(0, x) - \tilde{R}(0, x)B(x) - \tilde{R}(0, x)A(x)\tilde{R}(0, x), \quad (4.40a)$$

$$\frac{d\tilde{\tau}(0, x)}{dx} = -\tilde{\tau}(0, x)[B(x) + A(x)\tilde{R}(0, x)], \quad (4.40b)$$

$$\frac{d\tilde{T}(0, x)}{dx} = -[C(x) + \tilde{R}(0, x)A(x)]\tilde{T}(0, x), \quad (4.40c)$$

$$\frac{d\tilde{\mathscr{R}}(0, x)}{dx} = -\tilde{\tau}(0, x)A(x)\tilde{T}(0, x). \quad (4.40d)$$

These differential equations are valid for the selected flow $v(x)$ to the right, and $u(x)$ to the left, with x increasing in a positive direction.

The higher-order incremental coefficients are needed for the recursive equations; the coefficients for this particular flow can be derived by the same method as given later for the normal flow. Coefficients of fourth-order corrections are given in Appendix A.

6. INITIAL VALUE AND LINEAR TWO-POINT BOUNDARY-VALUE PROBLEMS

The derivations of the recursive equations and the differential equations for the transmission matrix coefficients have ignored the boundary conditions to be imposed on $u(x)$ and $v(x)$. The initial conditions for the transmission and reflection coefficients have been given, but they only serve to permit the computation of the coefficients. The functions $u(x)$ and $v(x)$ must also be assigned initial values at $x = 0$ or final values at $x = a$.

The type of boundary conditions to be assigned to the problem depend on the problem under consideration. If $u(0)$ and $v(0)$ are specified, the problem is an initial value problem. In some problems, $u(0)$ and $v(a)$ may be specified or $v(0)$ and $u(a)$ may be given. This type of problem, with one boundary condition specified at each end, is the two-point boundary-value problem.

Consider the initial value problem and the method of applying the reflection and transmission coefficients to compute $u(x)$ and $v(x)$. The differential equations for $u(x)$ and $v(x)$ are

$$\frac{du(x)}{dx} = B(x)u(x) + A(x)v(x), \qquad u(0) = C_1, \qquad (4.41\text{a})$$

$$-\frac{dv(x)}{dx} = D(x)u(x) + C(x)v(x), \qquad v(0) = C_2, \qquad (4.41\text{b})$$

subject to the given initial values. Functions $u(x)$ and $v(x)$ are computed from the equations

$$u(x) = \tau(0, x)u(0) + R(0, x)v(x), \qquad (4.6)$$

$$v(0) = \mathscr{R}(0, x)u(0) + T(0, x)v(x). \qquad (4.10)$$

Since $\tau(0, x)$, $R(0, x)$, $\mathscr{R}(0, x)$, and $T(0, x)$ can be computed directly from the recursive relations, the functions $u(x)$ and $v(x)$ can be constructed at each x.

The two-point boundary-value problem can be analyzed from knowledge of the transmission matrix, but not in a straightforward manner like the initial value problem. Assume that the two boundary conditions given are $u(0) = C_1$ and $v(a) = C_2$. The solution cannot be constructed solely from (4.6) and (4.10), since $v(0)$ is not known. The value of $v(0)$ can be determined by computing $\mathscr{R}(0, a)$ and $T(0, a)$. The constant $\mathscr{R}(0, a)$ cannot be computed without first finding $R(0, a)$, $T(0, a)$, and $\tau(0, a)$. Since the recursive equations are required for this computation, values of $R(0, x)$, $T(0, x)$, $\tau(0, x)$ and $\mathscr{R}(0, x)$ can be stored at selected values of x. Since the initial value $v(0)$ is given by

$$v(0) = \mathscr{R}(0, a)u(0) + T(0, a)v(a), \qquad (4.42)$$

then, upon substituting (4.42) into (4.6) and (4.10), $u(x)$ and $v(x)$ are

$$u(x) = \{\tau(0, x) + R(0, x)T^{-1}(0, x)[\mathscr{R}(0, a) - R(0, x)]\}u(0)$$
$$+ R(0, x)T^{-1}(0, x)T(0, a)v(a), \qquad (4.43\text{a})$$

$$v(x) = T^{-1}(0, x)[\mathscr{R}(0, a) - \mathscr{R}(0, x)]u(0) + T^{-1}(0, x)T(0, a)v(a). \qquad (4.43\text{b})$$

7. DIFFERENTIAL EQUATIONS FOR THE SOURCE FUNCTIONS

The recursive equations for the two source functions $g(0, x)$ and $h(0, x)$, where $u(x)$ and $v(x)$ are defined as

$$u(x) = \tau(0, x)u(0) + R(0, x)v(x) + g(0, x), \quad (3.31a)$$

$$v(0) = \mathscr{R}(0, x)u(0) + T(0, x)v(x) + h(0, x), \quad (3.31b)$$

were derived in Chapter 3. The differential equations for $g(0, x)$ and $h(0, x)$ are required for finding the incremental coefficients. To derive these equations, proceed in the same manner that was used to find the differential equations for the transmission matrix coefficients. From (3.31a),

$$\frac{du(x)}{dx} = \frac{d\tau(0, x)}{dx}u(0) + \frac{dR(0, x)}{dx}v(x) + R(0, x)\frac{dv(x)}{dx} + \frac{dg(0, x)}{dx}, \quad (4.44)$$

but

$$\frac{du(x)}{dx} = B(x)u(x) + A(x)v(x) + e(x)$$

and

$$-\frac{dv(x)}{dx} = D(x)u(x) + C(x)v(x) + f(x).$$

Substituting into (4.44) for $du(x)/dx$ and $dv(x)/dx$ and eliminating $u(x)$, the following equation for $dg(0, x)/dx$ is found.

$$\frac{dg(0, x)}{dx} = R(0, x)f(x) + [B(x) + R(0, x)D(x)]g(0, x) + e(x). \quad (4.45)$$

The initial condition, which must be placed on $g(0, 0)$, is obtained from (3.31a).

$$g(0, 0) = g(x, x) = 0.$$

The differential equation for $h(0, x)$ is obtained from (3.31b).

$$\frac{dh(0, x)}{dx} = T(0, x)[f(x) + D(x)g(0, x)]. \quad (4.46)$$

The initial condition for $h(0, x)$ is the same as $g(0, x)$.

$$h(0, 0) = h(x, x) = 0.$$

It should be brought to the attention of the reader that (4.45) and (4.46) are not dependent on $u(x)$ and $v(x)$; thus the inhomogeneous differential equations can be solved for linear initial value or two-point boundary conditions. The two-point boundary-value problem must be characterized by linear differential equations in the formulation above. Nonlinear differential equations are discussed in a later chapter.

8. THE INHOMOGENEOUS DIFFERENTIAL EQUATION AND THE MATRIX PRODUCT FORMULATION

The homogeneous and inhomogeneous differential equations have been considered when the recursive equations are to be used in the computational scheme. The inhomogeneous differential equation now needs to be investigated when the matrix product method is programmed for the solutions.

Suppose that the inhomogeneous differential equations of (1.1) and (1.2) are written as

$$\frac{dU}{dx} = \bar{A}(x)U(x) + \bar{f}(x), \qquad (4.47)$$

where

$$U(x) = \begin{bmatrix} u(x) \\ v(x) \end{bmatrix},$$

$$\bar{A}(x) = \begin{bmatrix} B(x) & A(x) \\ -D(x) & -C(x) \end{bmatrix},$$

and

$$\bar{f}(x) = \begin{bmatrix} e(x) \\ -f(x) \end{bmatrix}.$$

The inhomogeneous differential equations are now considered to be an initial value problem where

$$U(0) = \bar{C} = \begin{bmatrix} C_1 \\ C_2 \end{bmatrix},$$

with $\bar{A}(x)$ and $\bar{f}(x)$ continuous for $0 \le x \le a$.

Assume that a unique solution for $U(x)$ exists for $0 \le x \le a$, and let $Z(x)$

4 THE RICCATI, RECURSIVE, AND OTHER DIFFERENTIAL EQUATIONS

be a matrix solution of the equation

$$\frac{dZ(x)}{dx} = -Z(x)\bar{A}(x), \qquad (4.48)$$

where $Z(0) = I$ and $\bar{A}(x)$ is as given above. The reduced associated matrix equation corresponding to the homogeneous equation is

$$\frac{dU(x)}{dx} = \bar{A}(x)U(x). \qquad (4.49)$$

Equation (4.48) is the adjoint of (4.49) and, if (4.48) is multiplied by $U(x)$ and (4.49) by $Z(x)$, we find upon adding

$$U(x)\frac{dZ(x)}{dx} + \frac{dU(x)}{dx}Z(x) = Z(x)\bar{A}(x)U(x) - Z(x)\bar{A}(x)U(x)$$

or

$$\frac{d}{dx}[U(x)Z(x)] = 0.$$

Therefore $U(x)Z(x)$ must be a constant or $U(x)$ or $Z(x)$ must equal zero. If $Z(x)$ is nonsingular, $U(x)$ is given by

$$U(x) = Z^{-1}(x)\bar{C}. \qquad (4.50)$$

If $Z^{-1}(0) = I$, then $Z^{-1}(x)$ must be the principal matrix solution for (4.49), which is denoted by $\mathbf{Q}(0, x)$ hereafter. Assuming that $\mathbf{Q}^{-1}(0, x)$ is the principal matrix solution of (4.48), $\mathbf{Q}(0, x)$ is the fundamental matrix solution for the equations of interest. The equation

$$\frac{d\mathbf{Q}(0, x)}{dx} = \bar{A}(x)\mathbf{Q}(0, x) \qquad (4.51a)$$

and the adjoint equation

$$\frac{d\mathbf{Q}^{-1}(0, x)}{dx} = -\mathbf{Q}^{-1}(0, x)\bar{A}(x) \qquad (4.51b)$$

are the fundamental matrix solutions to the homogeneous differential equation.

To find the solution to the inhomogeneous differential equation, multiply (4.47) by $\mathbf{Q}^{-1}(0, x)$, (4.51b) by $U(x)$, and add the two equations together

$$\mathbf{Q}^{-1}(0, x)\frac{dU(x)}{dx} + \frac{d\mathbf{Q}^{-1}(0, x)}{dx}U(x)$$
$$= \mathbf{Q}^{-1}(0, x)\bar{A}(x)U(x) + \mathbf{Q}^{-1}(0, x)\bar{f}(x) - \mathbf{Q}^{-1}(0, x)\bar{A}(x)U(x),$$

or

$$\frac{d}{dx}[\mathbf{Q}^{-1}(0, x)U(x)] = \mathbf{Q}^{-1}(0, x)\bar{f}(x). \quad (4.52)$$

Integrating (4.52) by parts, over x, and making use of the initial condition, the desired equation specified through the fundamental matrix is found

$$U(x) = \mathbf{Q}(0, x)\bar{C} + \mathbf{Q}(0, x)\int_0^x \mathbf{Q}^{-1}(0, y)\bar{f}(y)\,dy. \quad (4.53)$$

Equation (4.53) is the solution to the inhomogeneous differential equation. The function $\mathbf{Q}(0, x)$ can be found from (3.29), where the fundamental matrix is defined for the medium extending from 0 to x. If $\bar{C}$ is specified and if $\bar{f}(x)$ is given, the integration involved in (4.53) can be carried out numerically.

9. INTEGRATION IN THE NEGATIVE x DIRECTIONS

The numerical integration has been concerned only with the thickness of the medium increasing in the direction of positive x, with $x = 0$ as the starting point of the integration. The equations derived are valid regardless of the starting point, but the recursive equations must be changed for integration in the negative x direction. The recursive equations were given for proceeding in the negative direction, but the differential equations have not been derived for this particular case.

The two-coupled equations,

$$\frac{du(x)}{dx} = B(x)u(x) + A(x)v(x), \quad (3.1a)$$

$$-\frac{dv(x)}{dx} = D(x)u(x) + C(x)v(x), \quad (3.1b)$$

are again of interest. Assume that the integration is to proceed in the direction of negative x but with $x > 0$. Now, by definition,

$$\frac{du(x)}{dx} = \frac{u(x) - u(x - \Delta x)}{\Delta x}, \quad (4.54)$$

as is usual for the derivative. A similar expression holds for $dv(x)/dx$. The function $u(x)$ is still taken in the direction of positive x, and $v(x)$ is assumed to be in the direction of negative x.

Now, from (3.1a) and (4.54), we have

$$u(x) = [I + B(x)\,\Delta x]u(x - \Delta x) + A(x)\,\Delta x v(x) \quad (4.55a)$$

4 THE RICCATI, RECURSIVE, AND OTHER DIFFERENTIAL EQUATIONS

and, from (3.1b),

$$v(x - \Delta x) = D(x)\,\Delta x u(x - \Delta x) + [I + C(x)\,\Delta x]v(x), \quad (4.55b)$$

where higher-order terms of Δx have been excluded.

Using the equations for $u(x)$ and $v(x - \Delta x)$,

$$u(x) = \tau(x - \Delta x, x)u(x - \Delta x) + R(x - \Delta x, x)v(x), \quad (4.56a)$$

$$v(x - \Delta x) = \mathcal{R}(x - \Delta x, x)u(x - \Delta x) + T(x - \Delta x, x)v(x), \quad (4.56b)$$

the incremental transmission matrix coefficients are obtained from (4.55). The signal flow graph for the problem differs from those given earlier. A

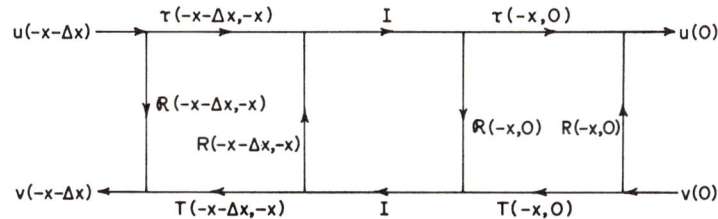

Figure 4.3. Integration in the direction of negative x.

section of thickness Δx must now be cascaded to the left of the medium under consideration as shown in Figure 4.3. The recursive equations for integration in the negative direction can now be written immediately as

$$\tau(x - \Delta x, a) = \tau(x, a)[I - R(x - \Delta x, x)\mathcal{R}(x, a)]^{-1}\tau(x - \Delta x, x), \quad (4.57a)$$

$$T(x - \Delta x, a) = T(x - \Delta x, x)[I - \mathcal{R}(x, a)R(x - \Delta x, x)]^{-1}T(x, a), \quad (4.57b)$$

$$R(x - \Delta x, a) = R(x, a) + \tau(x, a)[I - R(x - \Delta x, x)\mathcal{R}(x, a)]^{-1} \\ \cdot R(x - \Delta x, x)T(x, a), \quad (4.57c)$$

$$\mathcal{R}(x - \Delta x, a) = \mathcal{R}(x - \Delta x, x) + T(x - \Delta x, x) \\ \cdot [I - \mathcal{R}(x, a)R(x - \Delta x, x)]^{-1}\mathcal{R}(x, a)\tau(x - \Delta x, x). \quad (4.57d)$$

The reader should note that integration in the negative direction modifies the recursive equations in such a manner that $\mathcal{R}(x - \Delta x, a)$ generates the Riccati equation; this equation is the counterpart of $R(0, x + \Delta x)$ for positive integration.

The differential equations for the transmission matrix coefficients can be obtained from (4.55), (4.56), and (4.57). Consider (4.57d), where

$$\lim_{\Delta x \to 0}\left[\frac{\mathscr{R}(x - \Delta x, a) - \mathscr{R}(x, a)}{\Delta x}\right] = -\frac{d\mathscr{R}(x, a)}{dx}, \quad (4.58)$$

$$-\frac{d\mathscr{R}(x, a)}{dx} = \lim_{\Delta x \to 0}\left[\frac{\mathscr{R}(x - \Delta x, x)}{\Delta x} + \frac{T(x - \Delta x, x)}{\Delta x}\right.$$

$$\cdot [I - \mathscr{R}(x, a)R(x - \Delta x, x)]^{-1}$$

$$\left. \cdot \mathscr{R}(x, a)\tau(x - \Delta x, x) - \frac{\mathscr{R}(x, a)}{\Delta x}\right]; \quad (4.59)$$

then

$$-\frac{d\mathscr{R}(x, a)}{dx} = D(x) + C(x)\mathscr{R}(x, a) + \mathscr{R}(x, a)B(x) + \mathscr{R}(x, a)A(x)\mathscr{R}(x, a).$$

(4.60a)

It is assumed that the increment of integration is a positive number, that is, $|-\Delta x| = h$ is the increment of integration. The negative sign for Δx has been included in the differential equation. The remaining three differential equations follow from a similar process.

$$-\frac{d\tau(x, a)}{dx} = \tau(x, a)[B(x) + A(x)\mathscr{R}(x, a)], \quad (4.60\mathrm{b})$$

$$-\frac{dT(x, a)}{dx} = [C(x) + \mathscr{R}(x, a)A(x)]T(x, a), \quad (4.60\mathrm{c})$$

$$-\frac{dR(x, a)}{dx} = \tau(x, a)A(x)T(x, a). \quad (4.60\mathrm{d})$$

The recursive equations of (4.57) are useful in integrating in the direction of negative x. Such an integration procedure would hold if $u(a)$ and $v(a)$ were specified and the solutions for $0 < x < a$ were sought. In this particular case, the medium's thickness is increasing as x decreases.

It can be shown that, if the medium is of thickness a, extending from $x = 0$ to $x = a$, differential equations can be derived for a decreasing thickness medium. The medium's thickness is decreased by removing slabs on the left side of the medium, the right side of the medium being fixed at $x = a$. Redheffer [5] has derived these equations, and the reader should consult that paper for details. The recursive equations for the decreasing thickness can be derived from (4.57) by solving the equations for $\tau(x, a)$, $T(x, a)$, $R(x, a)$, and $\mathscr{R}(x, a)$.

4 THE RICCATI, RECURSIVE, AND OTHER DIFFERENTIAL EQUATIONS

10. THE INCREMENTAL COEFFICIENTS FOR THE T MATRIX

The incremental coefficients $R(x, x + \Delta x)$, $T(x, x + \Delta x)$, $\tau(x, x + \Delta x)$, and $\mathscr{R}(x, x + \Delta x)$ must be found before the recursive equations can be considered as useful for computation. Before giving one method of deriving these coefficients, the physical meaning of the incremental coefficients should be made clear. The incremental coefficients are a measure of the reflection and transmission coefficients for a slab of thickness Δx at position x imbedded in a semi-infinite medium. These coefficients are for the slab of thickness Δx in a matched medium with no other reflections occurring.

Since the slab of thickness Δx is considered to be located at x, the incremental coefficients should be defined in terms of $A(x)$, $B(x)$, $C(x)$, and $D(x)$ which characterize the medium. The incremental coefficients do not depend on the properties of the medium away from x since multireflections are assumed to be absent.

Consider a series expansion of $R(x, x + x)$ for a slab of thickness Δx.

$$R(x, x + \Delta x) = R(x, x) + \left.\frac{dR(x, s)}{ds}\right|_{s=x} \Delta x + \left.\frac{d^2 R(x, s)}{ds^2}\right|_{s=x} \frac{\Delta x^2}{2} + \cdots . \tag{4.61}$$

Now, $R(x, x)$ is the reflection coefficient for a slab of zero thickness located at x, which by definition is zero since it is equivalent to $R(0, 0)$. All terms appearing in (4.61) with $R(x, x)$ are zero. The first term on the right side in (4.61) is then zero. The higher-order terms in (4.61) can be found from the differential equation for $R(0, x)$.

$$\frac{dR(0, x)}{dx} = A(x) + B(x)R(0, x) + R(0, x)C(x) + R(0, x)D(x)R(0, x).$$

If the Riccati equation holds for $(0, x)$, then it also holds for (x, x) and, since $R(x, x) = 0$, then

$$\left.\frac{dR(x, s)}{ds}\right|_{s=x} = A(x). \tag{4.62}$$

Differentiating $dR(0, x)/dx$ once and with $R(x, x) = 0$,

$$\left.\frac{d^2 R(x, s)}{ds^2}\right|_{s=x} = A'(x) + B(x)A(x) + A(x)C(x). \tag{4.63}$$

Higher-order terms can be evaluated by continuing this process to the desired order.

Substituting (4.62) and (4.63) into (4.61), the second-order incremental coefficient for $R(x, x + \Delta x)$ is

$$R(x, x + \Delta x) = A(x)\Delta x + \left(\frac{dA(x)}{dx} + B(x)A(x) + A(x)C(x)\right)\frac{\Delta x^2}{2}.$$
(4.64a)

The three other incremental coefficients are evaluated by expanding the desired incremental coefficient in a Taylor series expansion and then using the appropriate differential equation. The remaining three incremental coefficients for a second-order expansion are given below.

$$T(x, x + \Delta x) = I + C(x)\Delta x + \left(\frac{dC(x)}{dx} + D(x)C(x) + C(x)C(x)\right)\frac{\Delta x^2}{2},$$
(4.64b)

$$\tau(x, x + \Delta x) = I + B(x)\Delta x + \left(\frac{dB(x)}{dx} + A(x)D(x) + B(x)B(x)\right)\frac{\Delta x^2}{2},$$
(4.64c)

$$\mathscr{R}(x, x + \Delta x) = D(x)\Delta x + \left(\frac{dD(x)}{dx} + C(x)D(x) + D(x)B(x)\right)\frac{\Delta x^2}{2}.$$
(4.64d)

The incremental coefficients of (4.64) are valid only for the medium's thickness increasing in the direction of positive x and with $u(x)$ in the direction of positive x. Fourth-order incremental coefficients have been found to give sufficient accuracy for most problems with a step size of $\Delta x = 0.01$.

The incremental coefficients can be written in a closed form in a few select problems. In general, this is not true and the truncated series must be used. If the truncated series is used, a compromise must be made between step size and the number of calculations which must be made. If the step size is large, errors result owing to the series approximation; whereas a small step size leads to computational error within the machine. A rule of thumb that the author has followed is to use a Δx such that $(\Delta x)^n$, where n is the order of the incremental coefficients, is set approximately equal to machine accuracy.

II. INCREMENTAL COEFFICIENTS FOR THE SOURCE FUNCTIONS

If the differential equations of (1.1) and (1.2) hold with $e(x) \neq 0$ and $f(x) \neq 0$, the recursive equations for $g(0, x + \Delta x)$ and $h(0, x + \Delta x)$ must be

4 THE RICCATI, RECURSIVE, AND OTHER DIFFERENTIAL EQUATIONS

involved in the computations. These source functions also have incremental coefficients which must be derived. The initial conditions $g(x, x) = h(x, x) = 0$ are valid for (4.45) and (4.46), which are employed to derive the pertinent incremental coefficients. Now, from (4.45),

$$\frac{dg(0, x)}{dx} = e(x) + R(0, x)f(x) + [B(x) + R(0, x)D(x)]g(0, x). \quad (4.45)$$

The Taylor series expansion of $g(x, x + \Delta x)$,

$$g(x, x + \Delta x) = g(x, x) + \frac{dg(x, s)}{ds}\bigg|_{s=x} \Delta x + \frac{d^2g(x, s)}{ds^2}\bigg|_{s=x} \frac{\Delta x^2}{2} + \cdots, \quad (4.65)$$

must be evaluated to the desired order. The first term of (4.65) is zero and $dg(x, s)/ds$ is given in (4.45)

$$\frac{dg(x, s)}{ds}\bigg|_{s=x} = e(x). \quad (4.66a)$$

The second-order term is obtained from the derivative of (4.45) with respect to s.

$$\frac{d^2g(x, s)}{ds^2}\bigg|_{s=x} = e'(x) + A(x)f(x) + B(x)e(x). \quad (4.66b)$$

The second-order incremental coefficient for the source function $g(0, x)$ is, therefore, given by the sum of (4.66a) and (4.66b) with proper multiplication by Δx.

$$g(x, x + \Delta x) = e(x) \Delta x + \left(\frac{de(x)}{dx} + A(x)f(x) + B(x)e(x)\right)\frac{\Delta x^2}{2}. \quad (4.67a)$$

Without giving the details which the reader can insert if necessary, the incremental coefficient for $h(0, x)$ is

$$h(x, x + \Delta x) = f(x) \Delta x + \left(\frac{df(x)}{dx} + D(x)e(x) + C(x)f(x)\right)\frac{\Delta x^2}{2}. \quad (4.67b)$$

Fourth-order incremental coefficients for $g(0, x)$ and $h(0, x)$ are given in Appendix A.

12. THE INCREMENTAL COEFFICIENTS FOR THE FUNDAMENTAL MATRIX: $u(x)$ IN DIRECTION OF $x+$, $x > 0$

The incremental coefficients for the fundamental matrix, the **Q** matrix, can be found in several ways. Two methods are described here: one obtained from the fundamental matrix itself, and the other derived by making use of the incremental coefficients for the transmission matrix. The fundamental matrix approach is outlined first.

The coupled differential equations, excluding the forcing functions, under consideration are

$$\frac{du(x)}{dx} = B(x)u(x) + A(x)v(x), \tag{3.1a}$$

$$\frac{dv(x)}{dx} = -D(x)u(x) - C(x)v(x), \tag{3.1b}$$

where the negative sign has now been assigned to $D(x)$ and $C(x)$, rather than to $dv(x)/dx$. Now let $u(x)$ and $v(x)$ be defined by the vector $U(x)$. The equations of (3.1) can then be written as

$$\frac{dU(x)}{dx} = \bar{A}(x)U(x), \tag{4.67}$$

where $\bar{A}(x)$ is the $\bar{n} \times \bar{n}$ matrix.

$$\bar{A}(x) = \begin{bmatrix} B(x) & A(x) \\ -D(x) & -C(x) \end{bmatrix}. \tag{4.68}$$

The solution of (4.67) is given by

$$U(x) = \mathbf{Q}(0, x)U(0),$$

where $\mathbf{Q}(0, x)$ is the fundamental matrix over the medium's thickness.

Assume that $\mathbf{Q}(0, x)$ is given by

$$\mathbf{Q}(0, x) = I + \int_0^x \bar{A}(y)\, dy + \int_0^x \bar{A}(y) \int_0^y \bar{A}(y_1)\, dy_1\, dy$$
$$+ \int_0^x \bar{A}(y) \int_0^y \bar{A}(y_1) \int_0^{y_1} \bar{A}(y_2)\, dy_2\, dy_1\, dy + \cdots. \tag{4.69}$$

Differentiating $Q(0, x)$ once with respect to x,

$$\frac{d\mathbf{Q}(0, x)}{dx} = \bar{A}(x) + \bar{A}(x) \int_0^x \bar{A}(y)\, dy + \bar{A}(x) \int_0^x \bar{A}(y) \int_0^y \bar{A}(y_1)\, dy\, dy + \cdots \tag{4.70}$$

or

$$\frac{d\mathbf{Q}(0, x)}{dx} = \bar{A}(x)\left[I + \int_0^x \bar{A}(y)\, dy + \int_0^x \bar{A}(y) \int_0^y \bar{A}(y_1)\, dy\, dy + \cdots \right]$$
$$= \bar{A}(x)\mathbf{Q}(0, x). \tag{4.71}$$

4 THE RICCATI, RECURSIVE, AND OTHER DIFFERENTIAL EQUATIONS

Differentiating $U(x)$ with respect to x,

$$\frac{dU(x)}{dx} = \frac{d\mathbf{Q}(0, x)}{dx} U(0),$$

but $d\mathbf{Q}(0, x)/dx$ is given by (4.71); thus

$$\frac{dU(x)}{dx} = \bar{A}(x)\mathbf{Q}(0, x)U(0) = \bar{A}(x)U(x), \quad (4.72)$$

which is the original differential equation for $U(x)$. The function $\mathbf{Q}(0, x)$ is the solution to (4.67), and this matrix is the fundamental matrix given earlier.

Equation (4.69) can now be utilized in the derivation for the incremental matrix of the fundamental matrix. The function $\mathbf{Q}(x, x + \Delta x)$ is by definition the incremental matrix which is now given by

$$\mathbf{Q}(x, x + \Delta x) = I + \int_{x}^{x+\Delta x} \bar{A}(y)\, dy + \int_{x}^{x+\Delta x} \bar{A}(y) \int_{y}^{y+\Delta y} \bar{A}(y_1)\, dy_1\, dy$$

$$+ \int_{x}^{x+\Delta x} \bar{A}(y) \int_{y}^{y+\Delta y} \bar{A}(y_1) \int_{y_1}^{y_1+\Delta y} \bar{A}(y_2)\, dy_2\, dy_1\, dy + \cdots .$$

(4.73)

Expanding (4.73) in a Taylor series,

$$\mathbf{Q}(x, x + \Delta x) = \mathbf{Q}(x, x) + \left.\frac{d\mathbf{Q}(x, s)}{ds}\right|_{s=x} \Delta x + \left.\frac{d^2\mathbf{Q}(x, s)}{ds^2}\right|_{s=x} \frac{\Delta x^2}{2} + \cdots .$$

(4.74)

Now, $\mathbf{Q}(x, x) = I$, since the integrals are zero over a zero range. The first derivative of $\mathbf{Q}(x, s)$ is given by (4.71), and the higher-order derivatives can be derived by continuing the process involved in (4.70) and (4.71). The incremental matrix is found by calculating the series representation to the desired order. The third-order incremental matrix is given below.

$$\mathbf{Q}(x, x + \Delta x) = I + \bar{A}(x)\Delta x + \left[\frac{d\bar{A}(x)}{dx} + \bar{A}(x)\bar{A}(x)\right]\frac{\Delta x^2}{2}$$

$$+ \left[\frac{d^2\bar{A}(x)}{dx^2} + \frac{2\, d\bar{A}(x)}{dx} + \bar{A}(x)\frac{d\bar{A}(x)}{dx} + \bar{A}(x)\bar{A}(x)\bar{A}(x)\right]\frac{\Delta x^3}{6}.$$

(4.75)

The incremental coefficients for the transmission matrix can be found from $\mathbf{Q}(x, x + \Delta x)$. The equations for the transmission matrix incremental coefficients are calculated from the equations below.

$$T(x, x + \Delta x) = Q_{22}^{-1}(x, x + \Delta x), \tag{4.76a}$$

$$R(x, x + \Delta x) = Q_{12}(x, x + \Delta x)Q_{22}^{-1}(x, x + \Delta x), \tag{4.76b}$$

$$\mathscr{R}(x, x + \Delta x) = -Q_{22}^{-1}(x, x + \Delta x)Q_{21}(x, x + \Delta x), \tag{4.76c}$$

$$\tau(x, x + \Delta x) = Q_{11}(x, x + \Delta x)$$
$$- Q_{12}(x, x + \Delta x)Q_{22}^{-1}(x, x + \Delta x)Q_{21}(x, x + \Delta x). \tag{4.76d}$$

The inverse process can be carried out; that is, given the incremental coefficients for the transmission matrix, the incremental coefficients for the fundamental matrix can be calculated. The pertinent equations are given below.

$$Q_{11}(x, x + \Delta x)$$
$$= \tau(x, x + \Delta x) - R(x, x + \Delta x)T^{-1}(x, x + \Delta x)\mathscr{R}(x, x + \Delta x), \tag{4.77a}$$

$$Q_{12}(x, x + \Delta x) = R(x, x + \Delta x)T^{-1}(x, x + \Delta x), \tag{4.77b}$$

$$Q_{21}(x, x + \Delta x) = -T^{-1}(x, x + \Delta x)\mathscr{R}(x, x + \Delta x), \tag{4.77c}$$

$$Q_{22}(x, x + \Delta x) = T^{-1}(x, x + \Delta x). \tag{4.77d}$$

The choice of which set of incremental coefficients to use rests with the individual working the problem. The author has not carried out an error analysis and therefore cannot advise which method is better for finding the incremental coefficients.

13. RECURSIVE EQUATIONS FOR THE INTEGRAL OF THE INHOMOGENEOUS DIFFERENTIAL EQUATION

The integration necessary in (4.53) for the inhomogeneous differential equation can be carried out by recursive equations. It is difficult to invert $\mathbf{Q}(0, x)$ in (4.53), and to find an integration scheme that is sufficiently accurate to solve for $u(x)$ and $v(x)$. The recursive equations for the homogeneous equation have been derived. Attention is now turned to the forcing function integral.

If forcing functions are present, $u(x)$ and $v(x)$ can be obtained from (3.31).

$$u(x) = \tau(0, x)u(0) + R(0, x)v(x) + g(0, x), \tag{3.31a}$$

$$v(0) = \mathscr{R}(0, x)u(0) + T(0, x)v(x) + h(0, x). \tag{3.31b}$$

4 THE RICCATI, RECURSIVE, AND OTHER DIFFERENTIAL EQUATIONS

Rearranging and making use of (3.25), $u(x)$ and $v(x)$ are given by

$$u(x) = Q_{11}(0, x)u(0) + Q_{12}(0, x)v(0)$$
$$+ g(0, x) - Q_{12}(0, x)h(0, x), \qquad (4.78a)$$
$$v(x) = Q_{21}(0, x)u(0) + Q_{22}(0, x)v(0) - Q_{22}(0, x)h(0, x), \qquad (4.78b)$$

or in matrix form as

$$\begin{bmatrix} u(x) \\ v(x) \end{bmatrix} = \begin{bmatrix} Q_{11}(0, x) & Q_{12}(0, x) \\ Q_{21}(0, x) & Q_{22}(0, x) \end{bmatrix} \begin{bmatrix} u(0) \\ v(0) \end{bmatrix} + \begin{bmatrix} G(0, x) \\ H(0, x) \end{bmatrix}, \qquad (4.79)$$

where

$$\begin{bmatrix} G(0, x) \\ H(0, x) \end{bmatrix} = \begin{bmatrix} 1 & -Q_{12}(0, x) \\ 0 & -Q_{22}(0, x) \end{bmatrix} \begin{bmatrix} g(0, x) \\ h(0, x) \end{bmatrix}. \qquad (4.80)$$

Writing the equations for $u(x + \Delta x)$ and $v(x + \Delta x)$, as functions of the incremental coefficients and $u(x)$ and $v(x)$, we find

$$\begin{bmatrix} u(x + \Delta x) \\ v(x + \Delta x) \end{bmatrix}$$
$$= \begin{bmatrix} Q_{11}(x, x + \Delta x) & Q_{12}(x, x + \Delta x) \\ Q_{21}(x, x + \Delta x) & Q_{22}(x, x + \Delta x) \end{bmatrix} \begin{bmatrix} u(x) \\ v(x) \end{bmatrix} + \begin{bmatrix} G(x, x + \Delta x) \\ H(x, x + \Delta x) \end{bmatrix}. \qquad (4.81)$$

Substituting for $u(x)$ and $v(x)$ in (4.81) from (4.80) and collecting terms, (4.81) can be written in terms of the initial conditions $u(0)$ and $v(0)$, $\mathbf{Q}(0, x)$, and past values of the forcing function. The forcing function vector on the right side of (4.81) is then

$$\begin{bmatrix} G(x, x + \Delta x) \\ H(x, x + \Delta x) \end{bmatrix} = \begin{bmatrix} 1 & -Q_{12}(x, x + \Delta x) \\ 0 & -Q_{22}(x, x + \Delta x) \end{bmatrix} \begin{bmatrix} g(x, x + \Delta x) \\ h(x, x + \Delta x) \end{bmatrix}. \qquad (4.82)$$

The equations for $u(x + \Delta x)$ and $v(x + \Delta x)$ are found to be

$$\begin{bmatrix} u(x + \Delta x) \\ v(x + \Delta x) \end{bmatrix} = \begin{bmatrix} Q_{11}(0, x + \Delta x) & Q_{12}(0, x + \Delta x) \\ Q_{21}(0, x + \Delta x) & Q_{22}(0, x + \Delta x) \end{bmatrix} \begin{bmatrix} u(0) \\ v(0) \end{bmatrix}$$
$$+ \begin{bmatrix} Q_{11}(x, x + \Delta x) & Q_{12}(x, x + \Delta x) \\ Q_{21}(x, x + \Delta x) & Q_{22}(x, x + \Delta x) \end{bmatrix} \begin{bmatrix} G(0, x) \\ H(0, x) \end{bmatrix}$$
$$+ \begin{bmatrix} 1 & -Q_{12}(x, x + \Delta x) \\ 0 & -Q_{22}(x, x + \Delta x) \end{bmatrix} \begin{bmatrix} g(x, x + \Delta x) \\ h(x, x + \Delta x) \end{bmatrix}. \qquad (4.83)$$

Comparing (4.83) and (4.79), the forcing function contributions are found directly from (4.83).

$$\begin{bmatrix} G(0, x + \Delta x) \\ H(0, x + \Delta x) \end{bmatrix} = \begin{bmatrix} Q_{11}(x, x + \Delta x) & Q_{12}(x, x + \Delta x) \\ Q_{21}(x, x + \Delta x) & Q_{22}(x, x + \Delta x) \end{bmatrix} \begin{bmatrix} G(0, x) \\ H(0, x) \end{bmatrix}$$
$$+ \begin{bmatrix} 1 & -Q_{12}(x, x + \Delta x) \\ 0 & -Q_{22}(x, x + \Delta x) \end{bmatrix} \begin{bmatrix} g(x, x + \Delta x) \\ h(x, x + \Delta x) \end{bmatrix}. \quad (4.84)$$

Equation (4.84) is thus the matrix recursive equation equivalent for the integral

$$\mathbf{Q}(0, x) \int_0^x \mathbf{Q}^{-1}(0, t) \begin{bmatrix} e(t) \\ -f(t) \end{bmatrix} dt = \begin{bmatrix} G(0, x) \\ H(0, x) \end{bmatrix}. \quad (4.85)$$

The incremental coefficients for the forcing functions $g(x, x + \Delta x)$ and $h(x, x + \Delta x)$ are given by (4.67).

The initial conditions for $G(x, x)$ and $H(x, x)$ can be derived from (4.81). If $x + \Delta x \to x$, then $\mathbf{Q}(x, x) = I$ and $G(x, x) = H(x, x) = 0$.

14. DIFFERENTIAL EQUATIONS WITH PERIODIC SOLUTIONS: INITIAL VALUE PROBLEMS

The continuous method presented in the last two chapters partially ignored the discrete model which can be useful in certain problems. If the functions $u(x)$ and $v(x)$ are periodic of a period ξ, then

$$u(x) = u(x + \xi) = u(x + n\xi) \quad (4.86)$$

and

$$v(x) = v(x + \xi) = v(x + n\xi). \quad (4.87)$$

Suppose now that the differential equations have been solved for the range $0 \leq x \leq x + \xi$ by using the continuous method. Assume also that the increment size and the incremental coefficients have been taken to a high order so that the accuracy is to the limits of the machine. All of the coefficients of the transmission matrix and the fundamental matrix are thus known.

The solution for any point $x + n\xi$ is then given by

$$u(x + n\xi) = \tau(0, x + n\xi)u(0) + R(0, x + n\xi)v(x + n\xi), \quad (4.88a)$$
$$v(0) = \mathscr{R}(0, x + n\xi)u(0) + T(0, x + n\xi)v(x + n\xi). \quad (4.88b)$$

As the composite coefficients of the transmission matrix can easily be found, the solution can be advanced from x to $x + n\xi$. The coefficients can be calculated from the discrete model.

$$R(0, x + n\xi) = R(x, x + n\xi) + \tau(x, x + n\xi)[I - R(0, x)\mathcal{R}(x, x + n\xi)]^{-1}$$
$$R(0, x)T(x, x + n\xi), \quad (4.89\text{a})$$
$$T(0, x + n\xi) = T(0, x)[I - \mathcal{R}(x, x + n\xi)R(0, x)]^{-1}T(x, x + n\xi). \quad (4.89\text{b})$$

The equations for $\mathcal{R}(0, x + n\xi)$ and $\tau(0, x + n\xi)$ can be written from the discrete model.

If the solutions are periodic, the coefficients for the thickness $(x, x + n\xi)$ can be written as

$$\begin{bmatrix} Q_{11}(x, x + n\xi) & Q_{12}(x, x + n\xi) \\ Q_{21}(x, x + n\xi) & Q_{22}(x, x + n\xi) \end{bmatrix} = \begin{bmatrix} Q_{11}(x, x + \xi) & Q_{12}(x, x + \xi) \\ Q_{21}(x, x + \xi) & Q_{22}(x, x + \xi) \end{bmatrix}^n, \quad (4.90)$$

where the matrix operation on the right is a matrix multiplication of order n. There is no need to carry out the incremental process, since the multiplication above is valid for the final n periods.

15. PROPERTIES OF THE FUNDAMENTAL MATRIX

There are several properties of the fundamental matrix which are important when working with the numerical methods given here. These properties are stated below.

PROPERTY 1. The fundamental matrix obeys the property

$$\mathbf{Q}(x, x) = \mathbf{Q}(0, 0) = I. \quad (4.91)$$

This property can easily be seen by noting that

$$\begin{bmatrix} u(x) \\ v(x) \end{bmatrix} = \mathbf{Q}(x, x) \begin{bmatrix} u(x) \\ v(x) \end{bmatrix}. \quad (4.92)$$

PROPERTY 2. The group property of the fundamental matrix is

$$\mathbf{Q}(x_0, x_2) = \mathbf{Q}(x_1, x_2)\mathbf{Q}(x_0, x_1). \quad (4.93)$$

This property can be verified by cascading the medium extending from x_0 to x_1 with a second medium extending from x_1 to x_2.

$$\begin{bmatrix} u(x_2) \\ v(x_2) \end{bmatrix} = \mathbf{Q}(x_0, x_2) \begin{bmatrix} u(x_0) \\ v(x_0) \end{bmatrix} = \mathbf{Q}(x_1, x_2)\mathbf{Q}(x_0, x_1) \begin{bmatrix} u(x_0) \\ v(x_0) \end{bmatrix}. \quad (4.94)$$

PROPERTY 3. This property is quite useful and is one that has not been discussed in the test. If a medium extends from x_0 to x_2 and has the fundamental matrix $\mathbf{Q}(x_0, x_2)$, then

$$\mathbf{Q}(x_2, x_0) = \mathbf{Q}^{-1}(x_0, x_2). \qquad (4.95)$$

This property implies that the inverse of the fundamental matrix calculated for x increasing from x_0 to x_2 is equivalent to the fundamental matrix $\mathbf{Q}(x_2, x_0)$ where the range has been interchanged.

If the original differential equations are

$$\frac{du(x)}{dx} = B(x)u(x) + A(x)v(x), \qquad (3.1a)$$

$$-\frac{dv(x)}{dx} = D(x)u(x) + C(x)v(x), \qquad (3.1b)$$

and they have the fundamental matrix $\mathbf{Q}(x_0, x_2)$, then the adjoint system

$$\frac{d\alpha(x)}{dx} = -\alpha(x)B(x) + \beta(x)D(x), \qquad (4.96a)$$

$$-\frac{d\beta(x)}{dx} = \alpha(x)A(x) - \beta(x)C(x), \qquad (4.96b)$$

will have the fundamental matrix $\mathbf{Q}^{-1}(x_0, x_2)$.

Further discussions of the differential equation and its adjoint can be found in Zadeh and Desoer [6]. There are certain problems that can be solved more readily by studying the two systems of differential equations rather than by looking at one system alone.

16. THE NUMERICAL SOLUTION OF ONE FIRST-ORDER DIFFERENTIAL EQUATION

It has been assumed in all of the discussion that two first-order differentials having the forms of (1.1) and (1.2) are to be solved. There are times when only one first-order differential equation is present; for example, suppose the differential equation

$$\dot{y}(t) = B(t)y(t) + e(t), \qquad y(0) = c_1 \qquad (4.97a)$$

is to be solved. It is a waste of computer time and programming effort to solve two equations when only the function $y(t)$ is desired. This would be the procedure to follow if the transmission matrix as defined were to be used.

The second equation which might be solved in addition to the equation above could be

$$\dot{z}(t) = C(t)z(t), \quad z(0) = c_2. \tag{4.97b}$$

Let $y(t) = u(x)$ and $z(t) = v(x)$, then we find the coupled equations

$$\frac{du(x)}{dx} = B(x)u(x) + e(x), \tag{4.98a}$$

$$-\frac{dv(x)}{dx} = C(x)v(x). \tag{4.98b}$$

If the incremental coefficients are considered [and observing that $A(x) = D(x) = 0$], it is not difficult to prove that $R(0, x) = \mathcal{R}(0, x) = 0$, and the solutions to (4.98a) and (4.98b) are

$$u(x) = \tau(0, x)u(0) + g(0, x), \tag{4.99a}$$
$$v(0) = T(0, x)v(x) + h(0, x). \tag{4.99b}$$

Since $R(0, x) = 0$, the recursive equations for $\tau(0, x + \Delta x)$ and $T(0, x + \Delta x)$ are independent:

$$\tau(0, x + \Delta x) = \tau(x, x + \Delta x)\tau(0, x), \tag{4.100a}$$
$$T(0, x + \Delta x) = T(0, x)\tau(x, x + \Delta x), \tag{4.100b}$$

and $g(0, x + \Delta x)$ and $h(0, x + \Delta x)$ are not coupled.

$$g(0, x + \Delta x) = g(0, x) + \tau(x, x + \Delta x)g(x, x + \Delta x), \tag{4.100c}$$
$$h(0, x + \Delta x) = h(0, x) + T(0, x)h(x, x + \Delta x). \tag{4.100d}$$

This means that (4.97a) can be solved by itself without considering a second equation. The solution for a first-order differential equation can, therefore, be found by finding $\tau(0, x)$ and $g(0, x)$ without computing the other elements of the transmission matrix.

17. SUMMARY

The differential equation equivalence to the recursive equations has been discussed in this chapter. It is shown that the right-hand reflection coefficient obeys the Riccati equation. The other five recursive equations, including the two source equations, obey linear differential equations with a functional dependency on the right-hand reflection coefficient.

The incremental coefficients for a medium can be found from a Taylor series expansion, and the "wave" behavior analyzed through the recursive equations. The fundamental matrix equations are described, and the relationship to the transmission matrix established.

REFERENCES

1. R. Bellman and R. Kalaba, *Quasilinearization and Nonlinear Boundary-Value Problems*, American Elsevier Co., New York, 1965.
2. W. T. Reid, Properties of Solutions of a Riccati Matrix Differential Equation, *J. Math. Mech.*, 9 (1960), 749.
3. R. Redheffer, The Riccati Equation: Initial Values and Inequalities, *Math. Ann.*, 133 (1957), 235.
4. P. Bailey and G. M. Wing, Some Recent Developments in Invariant Imbedding with Applications, *J. Math. Phys.*, 6 (1965), 453.
5. R. Redheffer, On Solutions of Riccati's Equation as Functions of the Initial Values, *J. Rat. Mech. Anal.*, 5 (1956), 835.
6. L. Zadeh and C. Desoer, *Linear Systems Theory: The State Space Approach*, McGraw-Hill Book Co., New York, 1963.

RELATED REFERENCES

R. Redheffer, Inequalities for a Matrix Riccati Equation, *J. Math. Mech.*, 8 (1959), 349.
R. Redheffer, The Mycielski-Paezkowski Diffusion Problem, *J. Math. Mech.*, 9 (1960), 607.
R. Redheffer, Supplementary Note on Matrix Riccati Equations, *J. Math. Mech.*, 9 (1960), 745.
W. Schmeidler, *Vortage Uber Determinaten und Martizen*, Akademie-Verlag, Berlin, 1949.
S. Twomey, H. Jacobowitz, and H. B. Howell, Matrix Methods for Multiple-Scattering Problems, *J. Atmos. Sci.*, 23 (1966), 289.

CHAPTER 5

Solution of Differential Equations: Initial Value Problems

The recursive equations for the transmission matrix and the fundamental matrix have been derived, and their relationship with the differential equations established. The analysis of coupled differential equations, such as encountered in coupled mode work, can be carried out using the "flow" concept and the recursive equations. To assure that the method gives suitable accuracy for numerical analysis, differential equations with known solutions have been studied. The numerical accuracy of the method is presented in this chapter.

There are many numerical schemes for solving differential equations and, although the method here adds to the ever growing number of methods, the invariant imbedding approach has some advantages not present in other methods. The Adams-Moulton, Runge-Kutta, or any of the well-known methods could be utilized in the work which follows. The writer recognizes the values of these methods and readily admits that the invariant imbedding method only makes the choice of which to use more difficult. The author does not personally feel that the researcher should abstain from using other methods. The researcher should approach each problem with an open mind and bring all the tools at his disposal to bear on the problem. This method is therefore viewed as an additional mathematical tool to place in the researcher's "tool box."

A first-order differential equation is solved initially to illustrate the method. As mentioned earlier, a second-order differential equation may be generated from the first-order equation to use the coupled mode method, or the first-order differential equation can be solved by calculating an incomplete transmission matrix. The necessary recursive equations are then programmed and the desired result obtained.

Higher-order differential equations can be solved by following the procedure used in the first-order equation. The second-order equation selected for illustrating the method is the Airy differential equation. The Airy functions have been tabulated by Smirnov [1], and his tables are used to check the results. Since the tables by Smirnov are given only for six places, the accuracy of the numerical method when applied to the Airy equation can be checked only to that order from the table.

In addition to the first- and second-order differential equations mentioned above, such problems as a nonlinear differential equation, a definite integral, and a matrix problem are discussed in this chapter. The work in this chapter is limited to obtaining numerical results in real space. The application of the method in the complex plane is given later. The question of comparable accuracy between the invariant imbedding scheme and other numerical methods is examined in a subsequent chapter.

I. FIRST-ORDER DIFFERENTIAL EQUATIONS

The invariant imbedding method is applied initially to a first-order differential equation which has a simple solution: the equation

$$\frac{du(x)}{dx} = u(x) \tag{5.1}$$

whose solution is

$$u(x) = e^x \tag{5.2}$$

where the initial condition $u(0) = 1$ holds. The particular equation has the property that the incremental coefficients for the transmission matrix can be written in a closed form.

The first step in the analysis is to find the coupled differential equation set. To find these equations, differentiate (5.1) with respect to x and let $du(x)/dx = v(x)$; then

$$\frac{d^2u(x)}{dx^2} = \frac{dv(x)}{dx} = v(x). \tag{5.3}$$

The coupled set of equations are

$$\frac{du(x)}{dx} = v(x) = B(x)u(x) + A(x)v(x), \tag{5.4a}$$

$$-\frac{dv(x)}{dx} = -v(x) = D(x)u(x) + C(x)v(x), \tag{5.4b}$$

from which the coefficients $A(x)$, $B(x)$, $C(x)$, and $D(x)$ are found when $u(x)$ is taken in the direction of positive x.

$$A(x) = 1, \qquad B(x) = 0, \qquad C(x) = -1, \qquad D(x) = 0.$$

5 SOLUTION OF DIFFERENTIAL EQUATIONS: INITIAL VALUE PROBLEMS

Since the functions are constants, all of the derivatives of $A(x)$, $B(x)$, $C(x)$, and $D(x)$ vanish. The incremental coefficients contain only the functions $A(x)$ and $C(x)$ and products of the two nonzero functions.

The incremental coefficients are obtained from the expansion given in Appendix A. The incremental coefficient $R(x, x + \Delta x)$ is

$$R(x, x + x) = \Delta x - \frac{\Delta x^2}{2} + \frac{\Delta x^3}{6} - \frac{\Delta x^4}{24} + \cdots, \tag{5.5}$$

where

$$R'(x, x) = A(x) = 1,$$
$$R''(x, x) = A(x)C(x) = -1,$$
$$R'''(x, x) = A(x)C(x)C(x) = 1,$$
$$R''''(x, x) = A(x)C(x)C(x)C(x) = -1.$$

The closed form for the incremental coefficient $R(x, x + \Delta x)$ is immediately recognized to be

$$R(x, x + \Delta x) = 1 - e^{-\Delta x}, \tag{5.6a}$$

where (5.5) represents the first four leading terms of the expansion of (5.6). The remaining incremental coefficients are

$$T(x, x + \Delta x) = 1 - \Delta x + \frac{\Delta x^2}{2} - \frac{\Delta x^3}{6} + \frac{\Delta x^4}{24} - \cdots$$
$$= e^{-\Delta x}, \tag{5.6b}$$
$$\mathscr{R}(x, x + \Delta x) = 0, \tag{5.6c}$$
$$\tau(x, x + \Delta x) = 1. \tag{5.6d}$$

The four incremental coefficients of (5.6) are the correct ones to use in the recursive relations for the transmission matrix formulation. Since this is an initial value problem, the fundamental matrix formulation can be used for finding $u(x)$ and $v(x)$. The incremental coefficients for the fundamental matrix approach can be obtained by substituting (5.6) into (4.77).

$$\mathbf{Q}(x, x + \Delta x) = \begin{bmatrix} 1 & (e^{\Delta x} - 1) \\ 0 & e^{\Delta x} \end{bmatrix}. \tag{5.7}$$

The initial conditions for the fundamental matrix are $Q_{11}(x, x) = Q_{22}(x, x) = I$, and $Q_{12}(x, x) = Q_{21}(x, x) = 0$. The solution of the initial value problem is then

$$\begin{bmatrix} u(x) \\ v(x) \end{bmatrix} = \begin{bmatrix} Q_{11}(0, x) & Q_{12}(0, x) \\ Q_{21}(0, x) & Q_{22}(0, x) \end{bmatrix} \begin{bmatrix} u(0) \\ v(0) \end{bmatrix}. \tag{5.8}$$

Since the incremental coefficients of (5.6) are in closed form, the step size selected is unimportant. Indeed, it can readily be seen that the correct answer to the problem is given by (5.7) and (5.8).

$$\begin{bmatrix} u(x) \\ v(x) \end{bmatrix} = \begin{bmatrix} 1 & (e^x - 1) \\ 0 & e^x \end{bmatrix} \begin{bmatrix} u(0) \\ v(0) \end{bmatrix}. \qquad (5.9)$$

The initial conditions for the coupled differential equations are $u(0) = v(0) = 1$; $u(x)$ is then obtained directly from (5.9).

$$u(x) = u(0) + (e^x - 1)v(0) = e^x. \qquad (5.10)$$

The differential equation above was programmed and the calculations carried out on an IBM 360 Model 44 machine using double precision software. The step size is, therefore, of no importance other than as a check on the accuracy of the internal arithmetic of the machine and the correctness of the formulation. The results of the machine computations are given in the Table 5.1 for various step sizes. $u(x)$ was calculated out to $x = 18$, but only the values at $x = 18$ are given. The digits which are in error have been underlined in the table. All of the computed values are correct to twelve places, beyond which the error increases as the number of calculations increases.

Table 5.1. Solution of $u'(x) = v(x)$ with Initial Value $u(0) = 1$

	$u(18) = 6.56599691373305114 \times 10^7$	
Δx	$u(x) \times 10^{-7}$	$v(x) \times 10^{-7}$
0.04	6.565996913731170	6.565996913731150
0.06	6.565996913731550	6.565996913731490
0.09	6.565996913732040	6.565996913732030
0.12	6.565996913732350	6.565996913732340
0.15	6.565996913732590	6.565996913732590
0.18	6.565996913732680	6.565996913732690
0.60	6.565996913732960	6.565996913732950
0.90	6.565996913732990	6.565996913733000
3.00	6.565996913733000	6.565996913733010
6.00	6.565996913733020	6.565996913733020
9.00	6.565996913733040	6.565996913733040

5 SOLUTION OF DIFFERENTIAL EQUATIONS: INITIAL VALUE PROBLEMS

The incremental coefficients have been calculated exactly in this case. The order of the correction made to the incremental coefficients is quite important in the computational scheme. A trade-off can be made between step size and the order of the incremental coefficients. The optimum situation arises when the incremental coefficients can be recognized as a particular infinite series. This is unlikely in most cases since it implies that the solution is known.

2. A NETWORK TRANSIENT PROBLEM

The previous section considered a differential equation which has incremental coefficients in a closed form. In this section a problem is discussed

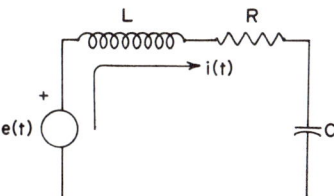

Figure 5.1. Simple network.

which does not have closed form incremental coefficients and where a forcing function is present. Consider the single-loop network given in Figure 5.1 with a unit step voltage excitation, $e(t) = 1$ for $t \geq 0$. The Kirchhoff loop equation for the network is

$$e(t) = L \frac{i(t)}{dt} + Ri(t) + \frac{1}{C} \int i(t)\, dt. \tag{5.11}$$

If the equation above is differentiated with respect to t, then

$$\delta(t) = L \frac{d^2 i(t)}{dt^2} + R \frac{di(t)}{dt} + \frac{i(t)}{C}, \tag{5.12}$$

where $\delta(t)$ is the impulse function, $\delta(0) = 1$, and $\delta(t) = 0$ for $t \neq 0$. Now let $u(t) = i(t)$ and $di(t)/dt = v(t)$, and the coupled differential equation set is found.

$$\frac{du(t)}{dt} = v(t) = B(t)u(t) + A(t)v(t), \tag{5.13a}$$

$$-\frac{dv(t)}{dt} = \frac{1}{LC} u(t) + \frac{R}{L} v(t) - \frac{\delta(t)}{L} = D(t)u(t) + C(t)v(t) + f(t). \tag{5.13b}$$

82   COUPLED MODES IN PLASMAS, ELASTIC MEDIA, PARAMETRIC AMPLIFIERS

The two equations of (5.13) are in the correct form and the solution is given by

$$\begin{bmatrix} u(t) \\ v(t) \end{bmatrix} = \begin{bmatrix} Q_{11}(0, t) & Q_{12}(0, t) \\ Q_{21}(0, t) & Q_{22}(0, t) \end{bmatrix} \begin{bmatrix} u(0-) \\ v(0-) \end{bmatrix} + \begin{bmatrix} Q_{11}(0, t) & Q_{12}(0, t) \\ Q_{21}(0, t) & Q_{22}(0, t) \end{bmatrix},$$

$$\int_0^t \begin{bmatrix} Q_{11}(0, s) & Q_{12}(0, s) \\ Q_{21}(0, s) & Q_{22}(0, s) \end{bmatrix}^{-1} \begin{bmatrix} e(s) \\ -f(s) \end{bmatrix} ds. \quad (5.14)$$

The circuit is initially at rest and therefore $u(0-)$ and $v(0-)$ are zero. The first term of (5.14) thus vanishes and only the second term remains. It is well known that, if the excitation is the unit impulse, then

$$\int_0^t \mathbf{Q}^{-1}(0, s) \delta(s - s_0) \, ds = \mathbf{Q}^{-1}(0, s_0). \quad (5.15)$$

Table 5.2. Transient Problem—Single Loop

R = 4 ohms,   L = 1 henry,   C = 0.16 farad,   Δt = 0.01

t	i (calc.)	i (corr.)
1.0	$-0.89997418 \times 10^{-1}$	$-0.89997511 \times 10^{-1}$
2.0	$-0.17233913 \times 10^{-2}$	$-0.17231354 \times 10^{-2}$
3.0	$0.16153342 \times 10^{-2}$	$0.16153699 \times 10^{-2}$
4.0	$0.62497138 \times 10^{-4}$	$0.62488971 \times 10^{-4}$
5.0	$-0.28388666 \times 10^{-4}$	$-0.28390088 \times 10^{-4}$
6.0	$-0.16882825 \times 10^{-5}$	$-0.16880956 \times 10^{-5}$
7.0	$0.48761943 \times 10^{-6}$	$0.48766152 \times 10^{-6}$
8.0	$0.40259111 \times 10^{-7}$	$0.40255555 \times 10^{-7}$
9.0	$-0.81599951 \times 10^{-8}$	$-0.81610803 \times 10^{-8}$
10.0	$-0.98361907 \times 10^{-9}$	$-0.89356209 \times 10^{-9}$

t	Δt = 0.001	i (corr.)
1.0	$-0.89997509 \times 10^{-1}$	$-0.89997511 \times 10^{-1}$
2.0	$-0.17231379 \times 10^{-2}$	$-0.17231354 \times 10^{-2}$
3.0	$0.16153695 \times 10^{-2}$	$0.16153699 \times 10^{-2}$
4.0	$0.62489050 \times 10^{-4}$	$0.62488971 \times 10^{-4}$
5.0	$-0.28390073 \times 10^{-4}$	$-0.28390088 \times 10^{-4}$
6.0	$-0.16880974 \times 10^{-5}$	$-0.16880956 \times 10^{-5}$
7.0	$0.48766107 \times 10^{-6}$	$0.48766152 \times 10^{-6}$
8.0	$0.40255584 \times 10^{-7}$	$0.40255551 \times 10^{-7}$
9.0	$-0.81610688 \times 10^{-8}$	$-0.81610803 \times 10^{-8}$
10.0	$-0.89356257 \times 10^{-9}$	$-0.89356209 \times 10^{-9}$

5 SOLUTION OF DIFFERENTIAL EQUATIONS: INITIAL VALUE PROBLEMS

The solution of (5.14) is found to be (where $s_0 = 0+$)

$$\begin{bmatrix} u(t) \\ v(t) \end{bmatrix} = \begin{bmatrix} Q_{11}(0, t) & Q_{12}(0, t) \\ Q_{21}(0, t) & Q_{21}(0, t) \end{bmatrix} \begin{bmatrix} Q_{12}(0, 0)/L \\ -Q_{11}(0, 0)/L \end{bmatrix}. \tag{5.16}$$

This single-loop circuit was solved by first finding $\mathbf{Q}(0, t)$ over the desired range. Since $Q_{12}(0, 0) = 0$ and $Q_{11}(0, 0) = 1$, the solution is easily constructed. The elements were chosen to be $R = 4$ ohms, $L = 1$ henry, and $C = 0.16$ farad. The calculations were carried out to $t = 10$ for several step sizes, all calculations were made with double precision arithmetic. The calculated and correct values are given in Table 5.2 for step sizes of 0.01 and 0.001.

3. THE AIRY DIFFERENTIAL EQUATION

The second-order differential equations

$$L(u) = u''(x) + [\lambda^2 x r_1(x) + q(x)]u(x) = 0 \tag{5.17}$$

and

$$L_\alpha(u) = u''(x) + [\lambda^2 x r_\alpha(x) + q(x)]u(x) = 0 \tag{5.18}$$

are interesting and useful in proving the accuracy of the computational method. The solutions of (5.17) and (5.18) are given by

$$u(x) = A(x)U_1[\phi(x), \alpha] + B(x)U_2[\phi(x), \alpha], \tag{5.19}$$

where $U_1[\phi(x), \alpha]$ and $U_2[\phi(x), \alpha]$ are the Airy functions. When $\alpha = 1$, the functions $U_1[s, \alpha]$ and $U_2[s, \alpha]$ are related to the Bessel functions of order $\pm \frac{1}{3}$.

$$U_1[s, 1] = \frac{\Gamma(\frac{2}{3})}{(3)^{1/3}} (s)^{1/2} J_{-1/3}(\tfrac{2}{3} s^{3/2}), \tag{5.20a}$$

$$U_2[s, 1] = (3)^{1/3} \Gamma(\tfrac{4}{3})(s)^{1/2} J_{1/3}(\tfrac{2}{3} s^{3/2}). \tag{5.20b}$$

The first derivatives of the Airy functions are defined in terms of the Bessel functions of order $\pm \frac{2}{3}$.

$$U_1'[s, 1] = \frac{-\Gamma(\frac{2}{3})}{(3)^{1/3}} s J_{2/3}(\tfrac{2}{3} s^{3/2}), \tag{5.21a}$$

$$U_2'[s, 1] = (3)^{1/3} \Gamma(\tfrac{4}{3}) s J_{-2/3}(\tfrac{2}{3} s^{3/2}). \tag{5.21b}$$

The differential equations of (5.18) and (5.19) are now solved using the numerical method and the computed values compared to the tabulated values of the Airy functions (see Smirnov [1]). In addition to serving as a comparative check on the numerical method, the method can be used to construct tables of Airy functions by assigning proper initial conditions.

Consider the equation given in (5.17) where $\alpha = 1$ and $q(x) = 0$. This particular equation is solved by the computational procedure presented in this book. Before proceeding with the numerical method, the general solution to (5.17) is investigated. The equation

$$u''(x) + \lambda^2 x r_1(x) u(x) = 0 \tag{5.22}$$

has the solution

$$u(x) = \frac{C_1}{\sqrt{w'(x)}} U_1[\lambda^{2/3} w(x), 1] + \frac{C_2}{\sqrt{w'(x)}} U_2[\lambda^{2/3} w(x), 1], \tag{5.23}$$

where

$$w(x) = \left(\tfrac{3}{2} \int_0^x \sqrt{t r_1(t)}\, dt \right)^{2/3}. \tag{5.24}$$

The first derivative of $u(x)$ is

$$u'(x) = C_1 \{ U_1'[\lambda^{2/3} w(x), 1] \lambda^{2/3} w'(x) - w'(x)^{-3/2} w''(x) U_1[\lambda^{2/3} w(x), 1] \}$$
$$+ C_2 \{ U_2'[\lambda^{2/3} w(x), 1] \lambda^{2/3} w'(x) - w'(x)^{-3/2} w''(x) U_2[\lambda^{2/3} w(x), 1] \}, \tag{5.25}$$

where the prime (') indicates differentiation with respect to x.

By selecting $r(x)$ such that $w''(x) = 0$, the Airy functions do not enter into (5.25); only the derivatives of U_1 and U_2 are present. If the initial conditions for the equations are properly chosen, the two Airy functions and the fractional-order Bessel functions can be generated directly from the coupled equations.

Let $u(0) = 0$ and $u'(0) = 1$; then

$$0 = \frac{C_1}{\sqrt{w'(0)}} U_1[\lambda^{2/3} w(0), 1] + \frac{C_2}{\sqrt{w'(0)}} U_2[\lambda^{2/3} w(0), 1] \tag{5.26}$$

and

$$1 = C_1 \sqrt{w'(0)}\, U_1'[\lambda^{2/3} w(0), 1] \lambda^{2/3} + C_2 \sqrt{w'(0)}\, U_2'[\lambda^{2/3} w(0), 1], \tag{5.27}$$

where it has been assumed that $w''(x) = 0$. The initial conditions on $U_1(s, 1)$, $U_2(s, 1)$, $U_1'(s, 1)$, and $U_2'(s, 1)$ are 1, 0, 0, and 1, respectively. Therefore, when $u(0) = 0$, $u'(0) = 1$ and $w'(0) = 1$,

$$C_1 = 0, \qquad C_2 = \lambda^{-2/3}.$$

The solutions for $u(x)$ and $u'(x)$ are therefore

$$u(x) = \lambda^{-2/3} U_2[\lambda^{2/3} w(x), 1] = Q_{12}(x) u'(0) \tag{5.28a}$$

and

$$u'(x) = U_2'[\lambda^{2/3} w(x), 1] = Q_{22}(x) u'(0) \tag{5.28b}$$

5 SOLUTION OF DIFFERENTIAL EQUATIONS: INITIAL VALUE PROBLEMS

when expressed in terms of the fundamental matrix. Under these conditions the Airy function of the second kind and the fractional-order Bessel function of order $-\frac{2}{3}$ are computed.

A similar analysis shows that the initial conditions $u(0) = 1$ and $u'(0) = 0$ lead to the generation of the Airy function of the first kind and the Bessel function of fractional order $+\frac{2}{3}$. By properly selecting the initial conditions, the two Airy functions and the Bessel functions of fractional order are generated.

The differential equation

$$u''(x) + \lambda^2 x u(x) = u''(x) + \frac{1}{3}xu(x) = 0 \tag{5.29}$$

satisfies the condition that $w''(x) = 0$. Now, from (5.24),

$$w(x) = \left(\frac{3}{2}\int_0^x \sqrt{t}\, dt\right)^{2/3} = x \tag{5.30}$$

and $w''(x) = 0$ as required.

The fundamental matrix can be computed independently of the initial conditions assigned to $u(0)$ and $u'(0)$. The solutions are found from

$$\begin{bmatrix} u(x) \\ u'(x) \end{bmatrix} = \begin{bmatrix} Q_{11}(0, x) & Q_{12}(0, x) \\ Q_{21}(0, x) & Q_{22}(0, x) \end{bmatrix} \begin{bmatrix} u(0) \\ u'(0) \end{bmatrix}, \tag{5.31}$$

where $u'(x) = v(x)$ in the coupled equations. Using the two initial conditions

(a) $u(0) = 0, \quad u'(0) = 1,$

and

(b) $u(0) = 1, \quad u'(0) = 0,$

the fundamental matrix gives the following functions.

$$Q_{11}(0, x) = U_1[\lambda^{2/3}x, 1], \tag{5.32a}$$

$$Q_{12}(0, x) = \lambda^{-2/3}U_2[\lambda^{2/3}x, 1], \tag{5.32b}$$

$$Q_{21}(0, x) = \lambda^{2/3}U_1'[\lambda^{2/3}x, 1], \tag{5.32c}$$

$$Q_{22}(0, x) = U_2'[\lambda^{2/3}x, 1]. \tag{5.32d}$$

The two Airy functions and the two Bessel functions are then given by the $\mathbf{Q}(0, x)$ matrix aside from the constant $\lambda^{\pm 2/3}$.

The computational results are given in Table 5.3. Fourth-order incremental coefficients were used in the computations which are made in double precision. A steep size of $\Delta x = 0.0000693$ was selected because of the presence of $\lambda^{2/3}$ in the arguments of U_1, U_2, etc. The computed values agree with the values

Table 5.3. Airy Functions

	Calculated		Smirnov	
s	$U_1(s, 1)$	$U_2(s, 1)$	$U_1(s, 1)$	$U_2(s, 1)$
0.0	1.000000000	0.000000000	1.00000	0.00000
1.0	0.838812310	0.918628888	0.83881	0.91863
2.0	−0.014978509	0.899179952	−0.01498	0.89918
3.0	−0.694729412	−0.510648971	−0.69473	−0.51065
4.0	0.219970335	−0.573222759	0.21997	−0.57322
5.0	0.381481898	0.831947798	0.38147	0.83194
6.0	−0.582829323	−0.472238214	−0.58283	−0.47224
7.0	0.498389412	0.028354180	0.49840	0.02837
8.0	−0.343568961	0.267644421	−0.34358	0.26764
9.0	0.233044545	−0.405190150	0.23305	−0.40518
10.0	−0.199194464	0.428719252	0.19920	0.42872

Derivatives

	Calculated		Smirnov	
s	$U_1'(s, 1)$	$U_2'(s, 1)$	$U_1'(s, 1)$	$U_2'(s, 1)$
0.0	0.000000000	1.000000000	0.00000	1.00000
1.0	−0.467354138	0.680336924	−0.46735	0.68034
2.0	−0.109740832	−0.883427883	−1.09741	−0.88343
3.0	0.106302236	−1.361273692	0.10630	−1.36128
4.0	1.208338819	1.397246074	1.20833	1.39725
5.0	−1.093729153	0.236117204	−1.09373	0.23612
6.0	0.173777598	−1.574964643	0.17378	−1.57496
7.0	0.6807159532	2.045190200	0.68069	2.04516
8.0	−1.187937321	−1.985206115	−1.18796	−1.98523
9.0	1.420738688	1.820813596	1.42074	1.82081
10.0	−1.500175554	−1.791444652	−1.50020	−1.79151

published in the tables of Smirnov for the first five significant figures. The accuracy beyond five figures is unknown, since Smirnov's tables are correct only to this extent.

4. INTEGRATION OF FUNCTIONS

The coupled mode method can be utilized in integrating a definite integral via the coupled differential equations. The integral is used to generate a set of coupled differential equations which are then solved in the manner described

5 SOLUTION OF DIFFERENTIAL EQUATIONS: INITIAL VALUE PROBLEMS

herein. Assume that the integral

$$u(x) = \int_a^x f(t)\,dt$$

must be solved. If a is a constant, then

$$\frac{du(x)}{dx} = f(x) \tag{5.33a}$$

and

$$\frac{d^2u(x)}{dx^2} = \frac{dv(x)}{dx} = \frac{df(x)}{dx}. \tag{5.33b}$$

The two differential equations of (5.33) are now in the coupled mode form and can be solved. Since (5.33) forms a coupled set, the calculated values of $u(x)$ and $u'(x) = v(x)$ are found.

To test the numerical method when applied to the integral, the error function was selected as a test problem. The error function is well-behaved for real arguments, and numerical tables are available for checking the accuracy. Consider the integral

$$H(x) = \frac{2}{\sqrt{\pi}} \int_0^x e^{-\alpha^2}\,d\alpha, \tag{5.34}$$

where $H(x)$ is the error function. To set up the coupled differential equations, the derivatives of (5.34) are taken.

$$H'(x) = \frac{2}{\sqrt{\pi}} e^{-x^2}, \qquad H(0) = 0, \tag{5.35a}$$

$$H''(x) = -2xH'(x) \qquad H'(0) = \frac{2}{\sqrt{\pi}}. \tag{5.35b}$$

Let $H(x) = u(x)$ and $H'(x) = v(x)$, then, from (5.35),

$$\frac{du(x)}{dx} = v(x) \tag{5.36a}$$

and

$$-\frac{dv(x)}{dx} = 2xv(x). \tag{5.36b}$$

The coefficients $A(x)$, $B(x)$, $C(x)$, and $D(x)$ are obtained from (5.36).

$$A(x) = 1, \qquad B(x) = 0, \qquad D(x) = 0, \qquad C(x) = 2x.$$

The fourth-order incremental coefficients are easily found by using the expansions given in Appendix A.

$$\mathbf{T}(x, x + \Delta x) = \begin{bmatrix} \tau(x, x + \Delta x) & R(x, x + \Delta x) \\ \mathcal{R}(x, x + \Delta x) & T(x, x + \Delta x) \end{bmatrix}$$

$$= \begin{bmatrix} 1 & 0 \\ 0 & 1 \end{bmatrix} + \begin{bmatrix} 0 & 1 \\ 0 & 2x \end{bmatrix}\Delta x + \begin{bmatrix} 0 & 2x \\ 0 & 2 + 4x^2 \end{bmatrix}\frac{\Delta x^2}{2}$$

$$+ \begin{bmatrix} 0 & 4x^2 \\ 0 & 4x + 8x^3 \end{bmatrix}\frac{\Delta x^3}{6} + \begin{bmatrix} 0 & 12x + 8x^3 \\ 0 & 12 + 32x^2 + 64x^4 \end{bmatrix}\frac{\Delta x^4}{24}.$$

This particular integral was solved as an initial value problem using the **Q** formulation. The computations were carried out on an IBM 360 Model 44 computer with double precision arithmetic. The step size selected was $\Delta x = 0.001$. The results of the calculations and the correct values are given in Table 5.4.

Table 5.4. Error Function

x	$H(x)$ (calc.)	$H(x)$ (corr.)
0.1	0.112462916018253	0.112462916018285
0.2	0.222702589210381	0.222702589210478
0.3	0.328262759459002	0.328628759459127
0.4	0.428392355046550	0.428392355046668
0.5	0.520499877812962	0.520499877813047
0.6	0.603856090847899	0.603856090847926
0.7	0.677801193837456	0.677801193837418
0.8	0.742100964707767	0.742100946707660
0.9	0.796908212423010	0.796908212422832
1.0	0.842700792949946	0.842700792949715

The underlined values in Table 5.4 disagree with the tabulated values and are assumed to be in error. It should be noted that the error is in the 13th digit; this is close to the limitation of double precision. The derivatives of $H(x)$ are not presented, but they agree quite closely with the tabulated values given in [2].

5. NONLINEAR DIFFERENTIAL EQUATIONS WITH INITIAL VALUES

Thus far only linear differential equations have been considered. To be useful in engineering work, the method should be capable of solving nonlinear equations. *Nonlinear* initial value problems can be solved by a simple extension of the procedure above for linear initial value problems.

Consider a second-order differential equation as given below.

$$\frac{d^2u(x)}{dx^2} + k(x)^\alpha u(x) = 0, \qquad (5.37)$$

where α may take on any value other than zero. This particular equation does not have the required form of the coupled equations. To obtain the correct form, let $u^\alpha(x)$ be decomposed in such a way as to obtain $u(x)$ separately and combine the residual function with $k(x)$. Let

$$u^\alpha = u^{\alpha-1}u;$$

then

$$\frac{du(x)}{dx} = v \qquad (5.38a)$$

and

$$-\frac{dv(x)}{dx} = k(x)[u(x)]^{\alpha-1}u(x) = k(u, x)u(x). \qquad (5.38b)$$

The equations are now correct in form with coefficients

$$A(u, v, x) = 1, \quad B(u, v, x) = 0, \quad C(u, v, x) = 0, \quad D(u, v, x) = k(x)[u(x)]^{\alpha-1}.$$

To solve the nonlinear initial value problem, proceed in the usual manner. The function $D(u, v, x)$ and the first three derivatives are needed to find the incremental coefficients. If

$$D(u, v, x) = k(x)[u(x)]^{\alpha-1}, \qquad (5.39a)$$

then

$$D'(u, v, x) = k'(x)[u(x)]^{\alpha-1} + k(x)(\alpha - 1)[u(x)]^{\alpha-1}v(x) \qquad (5.39b)$$

and

$$D''(u, v, x) = k''(x)[u(x)]^{\alpha-1} + 2k'(x)(\alpha - 1)[u(x)]^{\alpha-2}v(x)$$
$$+ k(x)(\alpha - 1)(\alpha - 2)[u(x)]^{\alpha-3}v^2(x)(\alpha - 1)[u(x)]^{2\alpha-2}v(x). \qquad (5.39c)$$

Higher-order derivatives can be found by continuing the procedure above. If

$u'(x)$ appears in the differentiation process, replace it by the value given in (5.38a). Similarly, $v'(x)$ is replaced by (5.38b).

The problem selected for checking the numerical procedure is a first-order equation

$$\frac{du(x)}{dx} = -2xu^2(x) \qquad (5.40a)$$

whose solution is

$$u(x) = (1 + x^2)^{-1}.$$

To solve this equation, find the second-order equation

$$\frac{d^2u(x)}{dx^2} = -2u^2(x) - 4xu(x)v(x) \qquad (5.40b)$$

which, along with the first-order equation of (5.40a), forms the coupled set. The four coefficients for the coupled set are recognized immediately.

$$A(u, v, x) = 0, \qquad B(u, v, x) = -2xu(x),$$
$$C(u, v, x) = 4xu(x), \qquad D(u, v, x) = 2u(x).$$

The incremental coefficients can now be determined using the expansion for $\mathbf{T}(x, x + \Delta x)$.

Equation (5.40a) was solved for various step sizes, and the computed value at $x = 18$ compared to the correct value given in (5.41). The formulation of the $\mathbf{Q}(0, x)$ matrix can be used since this is an initial value problem with initial conditions $u(0) = 1$ and $v(0) = 0$. The results of the calculations are presented in Table 5.5 along with the execution time for each step size. All calculations were carried out on an IBM Model 44 in double precision.

Table 5.5. Nonlinear Equation: $u'(x) = -2xu^2(x)$

$u(18) = 3.076923076923076 \times 10^{-3}$

Step Size	$u(18) \times 10^{-3}$ (calc.)	Time (sec)
0.005	3.076923076923316	43
0.01	3.076923076923512	21
0.04	3.076923076703242	6
0.06	3.076923074739142	4
0.09	3.076923057419246	3
0.12	3.076922987634689	2
0.15	3.076922789855592	2
0.18	3.076922336901019	2

5 SOLUTION OF DIFFERENTIAL EQUATIONS: INITIAL VALUE PROBLEMS

This particular differential equation has been investigated by Shanks [3] using higher-order Runge-Kutta numerical methods. The reader may find that paper of interest.

6. INHOMOGENEOUS DIFFERENTIAL EQUATIONS

The recursive equations for forcing functions were derived in Chapter 3. It is a simple task to solve an inhomogeneous differential equation by the method. The homogeneous equation is analyzed separately to obtain the transmission matrix; the forcing function contribution is then obtained using the transmission matrix and the forcing function. To illustrate the method, consider the two differential equations given below.

$$\frac{du(x)}{dx} = -3u(x) + v(x) + \exp(2x), \tag{5.41a}$$

$$-\frac{dv(x)}{dx} = u(x) + 5v(x) - \exp(x). \tag{5.41b}$$

The general solutions to the equations above are

$$u(x) = (C_1 + C_2 x)\exp(-4x) + \tfrac{7}{36}\exp(2x) + \tfrac{1}{25}\exp(x) \tag{5.42a}$$

and

$$v(x) = (C_2 - C_1 - C_2 x)\exp(-4x) - \tfrac{1}{36}\exp(2x) + \tfrac{4}{25}\exp(x). \tag{5.42b}$$

If the initial conditions $u(0) = 1$ and $v(0) = 0$ are selected, the arbitrary constants in (5.42) are

$$C_1 = \tfrac{689}{900}, \qquad C_2 = \tfrac{570}{900}.$$

To solve the coupled set with the forcing functions, first find the incremental coefficients of the transmission matrix from the homogeneous set.

$$\frac{du(x)}{dx} = -3u(x) + v(x),$$

$$-\frac{dv(x)}{dx} = u(x) + 5v(x).$$

The fourth-order incremental coefficients for $\mathbf{T}(x, x + \Delta x)$ are given in (5.43).

$$\mathbf{T}(x, x + \Delta x) = \begin{bmatrix} 1 & 0 \\ 0 & 1 \end{bmatrix} + \begin{bmatrix} -3 & 1 \\ 1 & 5 \end{bmatrix}\Delta x + \begin{bmatrix} 10 & 2 \\ 2 & 26 \end{bmatrix}\frac{\Delta x^2}{2}$$

$$+ \begin{bmatrix} -34 & 6 \\ 6 & 142 \end{bmatrix}\frac{\Delta x^3}{6} + \begin{bmatrix} 120 & 24 \\ 24 & 824 \end{bmatrix}\frac{\Delta x^4}{24}. \tag{5.43}$$

The forcing functions are
$$e(x) = \exp(2x),$$
$$f(x) = -\exp(x),$$
which are utilized to find $g(0, x)$ and $h(0, x)$ for calculating $u(x)$ and $v(x)$ from the equations of (3.31).

$$u(x) = \tau(0, x)u(0) + R(0, x)v(x) + g(0, x), \qquad (3.31a)$$
$$v(0) = \mathscr{R}(0, x)u(0) + T(0, x)v(x) + h(0, x). \qquad (3.31b)$$

The incremental coefficients for $g(x, x + \Delta x)$ and $h(x, x + \Delta x)$ must be known before the six recursive equations for the six functions of (3.31) can be calculated. The fourth-order incremental coefficients for $g(x, x + \Delta x)$ and $h(x, x + \Delta x)$ are given in Appendix A. Since the coefficients $A(x)$, $B(x)$, $C(x)$, and $D(x)$ are known, the incremental coefficients of $g(x, x + \Delta x)$ and $h(x, x + \Delta x)$ can be set up.

$$g(x, x + \Delta x) = \exp(2x)\,\Delta x - [\exp(2x) + \exp(x)]\frac{\Delta x^2}{2}$$
$$+ [9\exp(2x) - \exp(x)]\frac{\Delta x^3}{6}$$
$$- [16\exp(2x) + 15\exp(x)]\frac{\Delta x^4}{24}, \qquad (5.44a)$$

$$h(x, x + \Delta x) = -\exp(x)\,\Delta x + [\exp(2x) - 6\exp(x)]\frac{\Delta x^2}{2}$$
$$+ [13\exp(2x) - 38\exp(x)]\frac{\Delta x^3}{6}$$
$$+ [72\exp(2x) - 252\exp(x)]\frac{\Delta x^4}{24}. \qquad (5.44b)$$

The recursive equations for the transmission matrix and the forcing functions are now used to find $u(x)$ and $v(x)$. Since the problem above is an initial value one, $u(x)$ and $v(x)$ are computed from

$$\begin{bmatrix} u(x) \\ v(x) \end{bmatrix} = \mathbf{Q}(0, x)\begin{bmatrix} u(0) \\ v(0) \end{bmatrix} + \mathbf{Q}(0, x)\int_0^x \mathbf{Q}^{-1}(0, t)\begin{bmatrix} e(t) \\ -f(t) \end{bmatrix} dt. \qquad (5.45)$$

The two coupled mode equations of (5.41) were solved using double precision and the incremental coefficients given above. The exact answers were obtained from (5.42). The results of the computations are given in Table 5.6.

5 SOLUTION OF DIFFERENTIAL EQUATIONS: INITIAL VALUE PROBLEMS

Table 5.6. Inhomogeneous Coupled Equations

$u'(x) = -3u(x) + v(x) + \exp(2x); \quad -v'(x) = u(x) + 5v(x) - \exp(x)$

x	Function	Calculated	Correct
1.0	$u(x)$	1.57111372499786	1.57111372499808
1.0	$v(x)$	0.215651895145290	0.215651895144997
2.0	$u(x)$	10.9125509294531	10.9125509294508
2.0	$v(x)$	−0.334835577792391	−0.334835577812130
3.0	$u(x)$	79.2479254782423	79.2479254782175
3.0	$v(x)$	−7.99268195329957	−7.99268195348858
4.0	$u(x)$	581.814646074927	581.814646074726
4.0	$v(x)$	−74.0686848220344	−74.0686482354878
5.0	$u(x)$	4288.86043091568	4288.86043091407
5.0	$v(x)$	−588.100166616307	−588.100166627808

The calculations in Table 5.6 were made on an SDS Sigma 7 digital computer in double precision. The fundamental matrix approach was used with the recursive equations of Section 13, Chapter 4, to evaluate the integral.

7. A MATRIX DIFFERENTIAL EQUATION

The scalar version of the numerical method has been presented by examining several problems. To examine the matrix case, a two-loop network problem has been selected since the network transient solution can be solved by the transform method. The numerical method is of questionable value when the transform method is applicable; however, the problem serves as a check on the accuracy. The numerical method would be a powerful tool for nonlinear networks.

The transform method is difficult to use if the elements of the circuits are nonlinear; the numerical method presented does not become involved for nonlinear problems. The invariant imbedding method is valid for a select class of nonlinear problems. Furthermore one computer program, when properly written, can be utilized in a wide class of problems. The author is not advising that the Laplace transform method be ignored in the training of engineers but rather that the numerical method presented be used to supplement the transform method.

The multiloop network belongs to the class of problem in which coupling occurs, in this case coupling between mesh currents. If a multiloop network is analyzed by Kirchhoff's mesh law, the current in one loop is coupled to the current in other loops which share a common electrical element. If inductive coupling is permitted, common branches are no longer required to have current coupling. The problem to be discussed here ignores the inductive coupling, but this can easily be included.

Consider the two-loop network given in Figure 5.2, where all elements are assumed to be linear. The forcing function $e(t)$ may take on many forms,

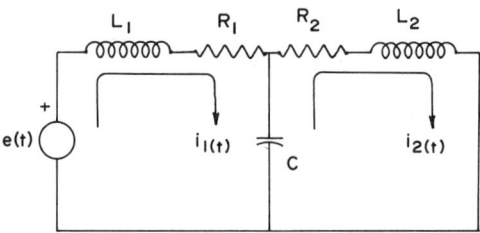

Figure 5.2. Two-loop network.

but for this analysis the unit step is selected. Now, by Kirchhoff's loop method,

$$e(t) = L_1 \frac{di_1(t)}{dt} + R_1 i_1(t) + \frac{1}{C} \int [i_1(t) - i_2(t)] \, dt, \qquad (5.46a)$$

$$0 = L_2 \frac{di_2(t)}{dt} + R_2 i_2(t) + \frac{1}{C} \int [i_2(t) - i_1(t)] \, dt. \qquad (5.46b)$$

Equations (5.46a) and (5.46b) can be put in the desired form by letting $i_1 = u_1$, $i_2 = u_2$, $di_1/dt = v_1$, and $di_2/dt = v_2$. Differentiating (5.46) with respect to t and making the above changes with respect to the variables, the four equations below are found.

$$\frac{du_1(t)}{dt} = v_1(t), \qquad (5.47a)$$

$$\frac{du_2(t)}{dt} = v_2(t), \qquad (5.47b)$$

$$-\frac{dv_1(t)}{dt} = \frac{u_1(t)}{L_1 C} - \frac{u_2(t)}{L_1 C} + \frac{R_1}{L_1} v_1(t) - \frac{1}{L_1} \frac{de(t)}{dt}, \qquad (5.47c)$$

$$-\frac{dv_2(t)}{dt} = -\frac{u_1(t)}{L_2 C} + \frac{u_2(t)}{L_2 C} + \frac{R_2}{L_2} v_2(t). \qquad (5.47d)$$

5 SOLUTION OF DIFFERENTIAL EQUATIONS: INITIAL VALUE PROBLEMS

If $e(t)$ is a unit step, then $de(t)/dt$ is a unit impulse, which is denoted by the usual function $\delta(t)$. The equations of (5.47) can now be written in matrix form.

$$\frac{du(t)}{dt} = Av(t), \tag{5.48a}$$

$$-\frac{dv(t)}{dt} = Du(t) + Cv(t) + f(t), \tag{5.48b}$$

where

$$A = I, \quad D = \begin{bmatrix} \frac{1}{L_1 C} & -\frac{1}{L_1 C} \\ -\frac{1}{L_2 C} & \frac{1}{L_2 C} \end{bmatrix}, \quad C = \begin{bmatrix} \frac{R_1}{L_1} & 0 \\ 0 & \frac{R_2}{L_2} \end{bmatrix},$$

$$B = 0 \quad f(t) = \begin{bmatrix} \frac{1}{L_1} \delta(t) \\ 0 \end{bmatrix}.$$

The equations of (5.48) are now in the desired form and the numerical method can be applied. The incremental coefficients $R(x, x + \Delta x)$, $T(x, x + \Delta x)$, $\tau(x, x + \Delta x)$, and $\mathcal{R}(x, x + \Delta x)$ can now be determined because the coefficients of the differential equations are known. To solve for the vectors $u(t)$ and $v(t)$, the recursive equations are solved over the desired range of t.

The transmission matrix or the fundamental matrix method can be used since this is an initial value problem. To find the currents at any particular time the initial conditions on $u(t)$ and $v(t)$ must be specified. For this problem let $u_1(0-) = u_2(0-) = du_1(0-)/dt = du_2(0-)/dt = 0$. Since $e(t)$ is a unit step, $de(t)/dt$ is a unit impulse and the initial conditions at $t = 0+$ are

$$u(0+) = v(0+) = e(0+) = \begin{bmatrix} 0 \\ 0 \end{bmatrix}, \quad f(0+) = \begin{bmatrix} -\delta(0+)/L_1 \\ 0 \end{bmatrix}.$$

These initial conditions are used in the **Q** formulation equation as given in (4.53).

$$U(t) = \mathbf{Q}(0, t)U(0-) + \mathbf{Q}(0, t) \int_0^t [\mathbf{Q}(0, 0)]^{-1} F(0+) \, dt. \tag{4.53}$$

Since $u(0-) = 0$ and $f(0-)$ is a unit impulse, (4.53) can be written as

$$U(t) = \mathbf{Q}(0, t)\mathbf{Q}^{-1}(0, 0+)F(0+) = \mathbf{Q}(0, t)F(0+), \tag{5.49}$$

where

$$U(t) = \begin{bmatrix} u(t) \\ v(t) \end{bmatrix}, \quad F(0+) = \begin{bmatrix} e(0+) \\ f(0+) \end{bmatrix}.$$

Equation (4.53) can be simplified to the form of (5.49) by recalling that the unit impulse gives

$$\int_{-\infty}^{\infty} g(y)\,\delta(y)\,dy = g(0)$$

for the integration over all time.

The selected problem can be solved by the transform method and the results used to check the accuracy of the numerical method. To find the fundamental matrix analytically, take the Laplace transform of (5.46) with undefined initial conditions.

$$sE(s) = (L_1 s^2 + R_1 s + C^{-1})I_1(s) - C^{-1}I_2(s) - L_1 s i_1(0)$$
$$- R_1 i_1(0) - L_1 \frac{di_1(0)}{dt}, \qquad (5.50\text{a})$$

$$0 = (L_2 s^2 + R_2 s + C^{-1})I_2(s) - C^{-1}I_1(s) - L_2 s i_2(0)$$
$$- R_2 i_2(0) - L_2 \frac{di_2(0)}{dt}. \qquad (5.50\text{b})$$

Solving for $I_1(s)$ and $I_2(s)$, we find

$$I_1(s) = Q_{11}(s)i_1(0) + Q_{12}(s)i_2(0) + Q_{13}(s)\frac{di_1(0)}{dt} + Q_{14}(s)\frac{di_2(0)}{dt}, \quad (5.51\text{a})$$

$$I_2(s) = Q_{21}(s)i_1(0) + Q_{22}(s)i_2(0) + Q_{23}(s)\frac{di_1(0)}{dt} + Q_{24}(s)\frac{di_2(0)}{dt}. \quad (5.51\text{b})$$

Multiplying (5.51a) and (5.51b) by s (this is equivalent to taking the time derivative), the remaining two equations for the fundamental matrix are found.

$$sI_1(s) = Q_{31}(s)i_1(0) + Q_{32}(s)i_2(0) + Q_{33}(s)\frac{di_1(0)}{dt} + Q_{34}(s)\frac{di_2(0)}{dt} \quad (5.51\text{c})$$

$$sI_2(s) = Q_{41}(s)i_1(0) + Q_{42}(s)i_2(0) + Q_{43}(s)\frac{di_1(0)}{dt} + Q_{44}(s)\frac{di_2(0)}{dt} \quad (5.51\text{d})$$

where $Q_{33}(s) = sQ_{11}(s)$ with similar expression for the other elements.

5 SOLUTION OF DIFFERENTIAL EQUATIONS: INITIAL VALUE PROBLEMS

$Q(s)$ is now given and $Q(0, t)$ can be found by taking the inverse Laplace transform of the coefficients of (5.51). The steady-state values can be determined by the final value theorem, which can be used to determine the limiting values of the currents and their derivatives. The limiting values may also be of interest in determining the limiting values for an infinitely thick medium.

The $Q(s)$ matrix of (5.51) is given below.

$$Q(s) = \frac{1}{(L_1s^2 + R_1s + C^{-1})(L_2s^2 + R_2s + C^{-1}) - C^{-2}}$$

$$\times \begin{bmatrix} (L_1s + R_1)(L_2s^2 + R_2s + C^{-1}) & C^{-1}(L_2s + R_2) \\ C^{-1}(L_1s + R_1) & (L_2s + R_2)(L_1s^2 + R_1s + C^{-1}) \\ s(L_1s + R_1)(L_2s^2 + R_2s + C^{-1}) & sC^{-1}(L_2s + R_2) \\ sC^{-1}(L_1s + R_1) & s(L_2s + R_2)(L_1s^2 + R_1s + C^{-1}) \end{bmatrix}$$

$$\begin{matrix} L_1(L_2s^2 + R_2s + C^{-1}) & L_2C^{-1} \\ L_1C^{-1} & L_2(L_1s^2 + R_1s + C^{-1}) \\ sL_1(L_2s^2 + R_2s + C^{-1}) & sL_2C^{-1} \\ sL_1C^{-1} & sL_2(L_1s^2 + R_1s + C^{-1}) \end{matrix} \quad . \quad (5.52)$$

To have a physical network, the circuit elements of (5.52) must all have positive values and the denominator of (5.52) must have roots which lie in the left half plane of the s domain with simple roots along the $j\omega$ axis. If $s = \sigma + j\omega$, then the roots for real s give exponentially decreasing functions for $i_1(t)$ and $i_2(t)$. When s is complex, the solutions are damped oscillatory functions. The currents $i_1(t)$ and $i_2(t)$, as well as the derivatives, must be bounded for the problem investigated here.

The network elements for the sample problem have been selected such that all of the roots of the denominator are real. The solutions for the current are therefore a sum of exponentially decreasing terms, that is, of the form $i_k(t) = \sum_{i=1}^{n} c_i \exp(p_i)$. The element values for the network are given in Table 5.7 along with the roots of the denominator of $Q(s)$.

Table 5.7. Network Elements and Roots

Network Elements					Roots			
R_1	R_2	L_1	L_2	C	p_1	p_2	p_3	p_4
5	4	1	1	0.5	0	$-3 + \sqrt{3}$	-3	$-3 - \sqrt{3}$

Now that the elements of the network have been fixed and the roots obtained, the currents $i_1(t)$ and $i_2(t)$ can be calculated. The inverse transform has been obtained by using the partial fraction expansion. The inverse transform of $\mathbf{Q}(s)$ can be written down by inspection. It should be pointed out that the inverse transform can be obtained by means of the numerical inverse method (see Bellman, Kalaba, and Lockett [4]). Their method is quite useful in system identification work (see Bellman, Kagiwada, and Kalaba [5] or Cook, Denman, and Carr [6]).

The matrix $\mathbf{Q}(0, t)$ is given below; these equations were used to calculate the correct numerical values for the desired variables.

$$Q_{11}(t) = \tfrac{5}{9} - \tfrac{2}{9}e^{-3t} - \frac{(2-\sqrt{3})(1+\sqrt{3})}{3(3+\sqrt{3})}e^{-(3+\sqrt{3})t}$$
$$- \frac{(2+\sqrt{3})(1-\sqrt{3})}{3(3-\sqrt{3})}e^{-(3-\sqrt{3})t}, \quad (5.53a)$$

$$Q_{12}(t) = \tfrac{4}{9} + \tfrac{2}{9}e^{-3t} - \frac{(1-\sqrt{3})}{3(3+\sqrt{3})}e^{-(3+\sqrt{3})t} - \frac{(1+\sqrt{3})}{3(3-\sqrt{3})}e^{-(3-\sqrt{3})t}, \quad (5.53b)$$

$$Q_{13}(t) = \tfrac{1}{9} - \tfrac{1}{9}e^{-3t} - \frac{(1+\sqrt{3})}{3(3+\sqrt{3})}e^{-(3+\sqrt{3})t} - \frac{(1-\sqrt{3})}{3(3-\sqrt{3})}e^{-(3-\sqrt{3})t}, \quad (5.53c)$$

$$Q_{14}(t) = \tfrac{1}{9} + \tfrac{2}{9}e^{-3t} - \frac{1}{3(3+\sqrt{3})}e^{-(3+\sqrt{3})t} - \frac{1}{3(3-\sqrt{3})}e^{-(3-\sqrt{3})t}, \quad (5.53d)$$

$$Q_{21}(t) = \tfrac{5}{9} + \tfrac{4}{9}e^{-3t} - \frac{(2-\sqrt{3})}{3(3+\sqrt{3})}e^{-(3+\sqrt{3})t} - \frac{(2+\sqrt{3})}{3(3-\sqrt{3})}e^{-(3-\sqrt{3})t}, \quad (5.53e)$$

$$Q_{22}(t) = \tfrac{4}{9} - \tfrac{4}{9}e^{-3t} + \frac{(1-\sqrt{3})^2}{6(3+\sqrt{3})}e^{-(3+\sqrt{3})t} + \frac{(1+\sqrt{3})^2}{6(3-\sqrt{3})}e^{-(3-\sqrt{3})t}, \quad (5.53f)$$

$$Q_{23}(t) = \tfrac{1}{9} + \tfrac{2}{9}e^{-3t} - \frac{1}{3(3+\sqrt{3})}e^{-(3+\sqrt{3})t} - \frac{1}{3(3-\sqrt{3})}e^{-(3-\sqrt{3})t}, \quad (5.53g)$$

$$Q_{24}(t) = \tfrac{1}{9} - \tfrac{4}{9}e^{-3t} + \frac{(1-\sqrt{3})}{6(3+\sqrt{3})}e^{-(3+\sqrt{3})t} + \frac{(1+\sqrt{3})}{6(3-\sqrt{3})}e^{-(3-\sqrt{3})t}. \quad (5.53h)$$

The remaining eight terms can be obtained from (5.53) by noting that

$$Q_{3j}(t) = \frac{d}{dt}Q_{1j}(t) \qquad (5.54)$$

5 SOLUTION OF DIFFERENTIAL EQUATIONS: INITIAL VALUE PROBLEMS

and

$$Q_{4j}(t) = \frac{d}{dt} Q_{2j}(t). \tag{5.55}$$

The correct values of $i_1(t)$, $i_2(t)$, $di_1(t)/dt$, and $di_2(t)/dt$ are obtained from (5.49); $Q(0, t)$ is as given in (5.53), (5.54), and (5.55). The argument $(0, t)$ has been replaced by (t) in the expressions above.

Table 5.8. Matrix Equation

Q_{ij}	$t = 5\pi$ (calc.)	$t = 5\pi$ (corr.)
Q_{11}	0.555556	0.555556
Q_{12}	0.444444	0.444444
Q_{13}	0.111111	0.111111
Q_{14}	0.111111	0.111111
Q_{21}	0.555556	0.555556
Q_{22}	0.444444	0.444444
Q_{23}	0.111111	0.111111
Q_{24}	0.111111	0.111111
Q_{31}	-0.203969×10^{-8}	-0.203971×10^{-8}
Q_{32}	0.203969×10^{-8}	0.203972×10^{-8}
Q_{33}	-0.546533×10^{-9}	-0.546539×10^{-9}
Q_{34}	0.746578×10^{-9}	0.746586×10^{-9}
Q_{41}	0.278626×10^{-8}	0.278630×10^{-9}
Q_{42}	-0.278627×10^{-8}	-0.278630×10^{-9}
Q_{43}	0.746578×10^{-9}	0.746586×10^{-9}
Q_{44}	-0.101984×10^{-8}	-0.101986×10^{-8}

The $Q(0, t)$ matrix was calculated numerically by the method described in the previous chapters with the coefficients given in (5.47). All calculations were made in double precision on the IBM 360 Model 44 machine. A step size of $\Delta t = \pi/200$ was selected, and the incremental coefficients were taken to fourth order. The results are given in Table 5.8, where the calculated and correct values from (5.53), (5.54), and (5.55) are given for comparison. The values of the sixteen elements of $Q(0, t)$ are given at $t = 5\pi$.

The recursive equations for the forcing functions were not required in this problem owing to the simple form of the excitation. The decision concerning which formulation, the fundamental matrix or the forcing function recursive equations, depends on the type of forcing function. It is a waste of computational time to use the forcing function recursive equations for an impulse function. The forcing functions $g(0, x)$, $h(0, x)$, $G(0, x)$, and $H(0, x)$ have values for the first step and there is no contribution for $t > 0+$.

8. NUMERICAL CALCULATIONS OF EIGENVALUES

The numerical method presented can be utilized in computing eigenvalues; one example is presented here. Assume that the differential equation

$$\frac{d^2u(x)}{dx^2} + \lambda^2 u(x) = 0 \tag{5.56}$$

is to be investigated and the eigenvalues calculated. The eigenvalues are values of λ such that $u(0) = u(x_i) = 0$, where $0 < x_i \leq a$. The length x_i is the characteristic length for the differential equation, and it is the length of the class $0 < x_1 < x_2 < x_3 \cdots$ for which a nontrivial solution exists.

If the fundamental matrix holds, then

$$u(x) = Q_{11}(0, x)u(0) + Q_{12}(0, x)v(0),$$
$$v(x) = Q_{21}(0, x)u(0) + Q_{22}(0, x)v(0).$$

In order for $u(x_i)$ to be zero, $Q_{12}(0, x) = 0$ for the critical length x_i. The procedure is therefore to carry out the numerical calculations for a fixed λ until $Q_{12}(0, x_i) = 0$. This gives the critical length for that value of λ. Conversely, if x_i is a fixed value for which λ_i is sought, the search must be made over λ_i as a variable.

To test the numerical accuracy, a value of $\lambda = \pi$ was selected and the first four values of x_i were calculated. The results are presented in Table 5.9.

Table 5.9. Tabulated Critical Lengths

x_i-	x_i+	$Q_{12}-$	$Q_{12}+$
0.99998645	1.00001061	0.1356×10^{-4}	-0.1059×10^{-4}
0.99999715	2.00001088	-0.2869×10^{-5}	0.1086×10^{-4}
2.99999900	3.00000859	0.1023×10^{-5}	-0.857×10^{-5}
3.99999486	4.00000223	-0.5159×10^{-5}	0.2212×10^{-5}

Since the medium is discrete, the limit on the accuracy depends on the limiting value of Δx. The results are therefore given with bounds on x_i- and x_i+ and $Q_{12}-$ and $Q_{12}+$ for the corresponding values of x_i.

9. SUMMARY

Numerous examples have been discussed and the computational results presented in this chapter. The writer has tried to present the details which are encountered in using the method so that the reader can apply the method. Some readers may find that much of the material could have been condensed and left as an exercise. In spite of the criticisms which may be expressed by reviewers of this book, the author has elected to include the details.

The computational accuracy for the problems selected has been adequate for most engineering problems. The accuracy can be increased where necessary by calculating higher-order incremental coefficients. Only fourth-order incremental coefficients were used in the presented problems. An attempt has been made to select problems that present representative applications to the reader so he can master the method. A reviewer of the book may argue that the examples do not reveal the true computational difficulties with a numerical method. This may be true, and such shortcomings are revealed as the method is applied to an assortment of equations. Time is required for the critical examination which this or any method must undergo.

Although the differential equations discussed in this chapter are not coupled mode equations in the true sense, the equations have been selected to verify the feasibility of the method. One of the problems encountered in solving coupled mode problems of elastic and electromagnetic waves is that so few exact solutions are known. The author would have devoted more time to treating coupled mode problems had numerical values been available for comparison.

REFERENCES

1. A. D. Smirnov, *Tables of Airy Functions and Special Confluent Hypergeometric Functions*, Pergamon Press, New York, 1960.
2. N.B.S. Staff, *Tables of the Error Function and Its Derivatives*, National Bureau of Standards, Applied Math. Series No. 41 (1954).
3. E. B. Shanks, Solutions of Differential Equations by Evaluation of Functions, *Math. of Comp.*, 20 (1966), 21.
4. R. Bellman, R. Kalaba, and J. Lockett, *Numerical Inversion of the Laplace Transform*, American Elsevier Co., New York, 1965.
5. R. Bellman, H. Kagiwada, and R. Kalaba, Identification of Linear Systems via Numerical Inversion of Laplace Transforms, *IEEE Trans. of Automatic Controls*, AC-10 (1965), 111.
6. G. Cook, E. Denman, and H. Carr, Numerical Inversion of Laplace Transforms by the Laguerre-Gauss Quadrature Approximation, *IEEE Trans. of Automatic Controls*, AC-12 (1967), 623.

CHAPTER 6

Two-Point Boundary-Value Problems

There is sufficient interest in linear and nonlinear two-point boundary-problems to justify the inclusion of some material on these types of problems. The linear two-point boundary-value problem can be solved using the recursive method given earlier by making two passes through the medium. The nonlinear two-point boundary-value problem requires an iterative scheme to obtain the transmission matrix elements, since the values of $u(x)$ and $v(x)$ cannot be calculated accurately in a single pass. The method of quasilinearization, described in the book by Bellman and Kalaba, is valuable in linearizing the equations. Once the equations are linearized, an iterative method makes it possible to calculate the transmission matrix.

The linear two-point boundary-value problem is discussed first. If the coupled differential equations are linear, the transmission matrix does not involve $u(x)$ and $v(x)$. The reflection and transmission elements can then be calculated. This is a rather simple problem even when forcing functions are present.

I. LINEAR TWO-POINT BOUNDARY-VALUE PROBLEM

The differential equations

$$\frac{du}{dx} = B(x)u(x) + A(x)v(x) + e(x), \quad u(0) = C_1, \quad (1.1)$$

$$-\frac{dv}{dx} = D(x)u(x) + C(x)v(x) + f(x), \quad v(a) = C_2, \quad (1.2)$$

are to be solved with the specified boundary conditions. Since the initial value of $v(0)$ is missing, the functions $u(x)$ and $v(x)$ cannot be constructed on the first pass through the medium. To find $u(x)$ and $v(x)$, the missing value of $v(0)$ must be found.

Since the differential equations are linear, $R(0, x)$, $T(0, x)$, $\mathcal{R}(0, x)$, and $\tau(0, x)$ depend only on $A(x)$, $B(x)$, $C(x)$, $D(x)$, and x. These elements are first calculated for $x = a$ by walking through the medium in a recursive fashion. If $u(x)$ and $v(x)$ obey the equations

$$u(x) = \tau(0, x) + R(0, x)v(x) + g(0, x), \qquad (3.31a)$$
$$v(0) = \mathscr{R}(0, x)u(0) + T(0, x)v(x) + h(0, x), \qquad (3.31b)$$

knowledge of $R(0, x)$, $T(0, x)$, $\mathscr{R}(0, x)$, and $\tau(0, x)$ is not sufficient to find $v(0)$. The two functions $g(0, x)$ and $h(0, x)$ must also be known for $x = a$. If $u(0)$, $v(a)$, $R(0, a)$, $T(0, a)$, $\mathscr{R}(0, a)$, $\tau(0, a)$, $g(0, a)$, and $h(0, a)$ are known, $u(a)$ and $v(0)$ can be found.

Assuming that the equations are linear and $e(x)$ and $f(x)$ are not functions of $u(x)$ and $v(x)$, $g(0, x)$ and $h(0, x)$ can be calculated from the recursive equations for these functions; see (3.33a) and (3.33b). These functions are calculated simultaneously with the transmission matrix. Only the final values of $g(0, x)$, $h(0, x)$, and the transmission matrix are required to find $v(0)$. All of the intermediate values of $g(0, x)$, $h(0, x)$, and the transmission matrix are discarded on the first pass through the medium. Once $v(0)$ is found, the recursive procedure is restarted at $x = 0$ and the functions $u(x)$ and $v(x)$ are found. This procedure reduces the storage requirements and requires about the same amount of time to construct the solutions as when the intermediate values are stored.

The fundamental matrix formulation can also be used in solving the two-point boundary-value problem. The recursive equations for the fundamental matrix are used with the same two-pass procedure outlined above.

2. MIXED LINEAR TWO-POINT BOUNDARY-VALUE PROBLEMS

Occasionally, problems may occur in which mixed boundary conditions are given. That is, suppose that $u(0)$ is an m-order vector and $v(a)$ is an m-order vector. If $u(0)$ is specified for $m - p$ values, then $v(a)$ is specified for $m + p$ values. This set of boundary conditions can be handled in the same way as described in the previous section. Consider a system of four equations with forcing functions, where

$$u(x) = \begin{bmatrix} u_1(x) \\ u_2(x) \end{bmatrix}, \quad v(x) = \begin{bmatrix} v_1(x) \\ v_2(x) \end{bmatrix} \qquad (6.1)$$

with $u_1(0) = c_1$, $u_2(a) = c_2$, $v_1(a) = c_3$, and $v_2(a) = c_4$.

The functions $u_2(0)$, $v_1(0)$, and $v_2(0)$ must be found before the solutions $u(x)$ and $v(x)$ are known. To find the missing initial conditions, proceed as before. Calculate the transmission matrix and the contribution due to the forcing terms. The missing initial conditions are then found from (3.31a) and (3.31b). The equations can be rearranged to obtain a suitable form to find the three missing initial values.

3. QUASILINEARIZATION OF NONLINEAR DIFFERENTIAL EQUATIONS

The most difficult type of two-point boundary-value problem to solve is that for which the differential equations are nonlinear. If the differential equations are nonlinear, the transmission matrix is functionally dependent on $u(x)$ and $v(x)$. The only way to eliminate this functional dependency is to quasilinearize the differential equations and then solve the equations by iteration. The mathematical theory of quasilinearization is not discussed; only the application to linearizing the equations is described. It is suggested that the reader consult the book by Bellman and Kalaba for unanswered questions.

Assume now that the differential equations of (1.1) and (1.2) are nonlinear, where

$$\frac{du}{dx} = f(u, v, x) + e(x), \tag{6.2a}$$

$$-\frac{dv}{dx} = g(u, v, x) + f(x). \tag{6.2b}$$

The transmission matrix or the fundamental matrix is found from the homogeneous equations; thus $e(x)$ and $f(x)$ are ignored temporarily in the discussion. To linearize (6.2a) and (6.2b), expand the equations in a Taylor series with $u(x)$ and $v(x)$ as variables. Assume that $u_0(x)$ and $v_0(x)$ are initial approximations to $u(x)$ and $v(x)$. The two variable expansions are then

$$f(u, v, x) = f(u_0, v_0, x) + \frac{\partial f}{\partial u}(u, v, x)\bigg|_{u_0, v_0}(u - u_0)$$
$$+ \frac{\partial f}{\partial v}(u, v, x)\bigg|_{u_0, v_0}(v - v_0) + \cdots, \tag{6.3a}$$

$$g(u, v, x) = g(u_0, v_0, x) + \frac{\partial g}{\partial u}(u, v, x)\bigg|_{u_0, v_0}(u - u_0)$$
$$+ \frac{\partial g}{\partial v}(u, v, x)\bigg|_{u_0, v_0}(v - v_0) + \cdots, \tag{6.3b}$$

6 TWO-POINT BOUNDARY-VALUE PROBLEMS

where higher-order terms are neglected. The homogeneous part of (6.2a) and (6.2b) can now be written as

$$\frac{du(x)}{dx} = f(u_0, v_0, x) + \frac{\partial f}{\partial u}(u, v, x)\bigg|_{u_0, v_0} (u - u_0)$$
$$+ \frac{\partial f}{\partial v}(u, v, x)\bigg|_{u_0, v_0} (v - v_0) \qquad (6.4a)$$

and

$$-\frac{dv(x)}{dx} = g(v_0, v_0, x) + \frac{\partial g}{\partial u}(u, v, x)\bigg|_{u_0, v_0} (u - u_0)$$
$$+ \frac{\partial g}{\partial v}(u, v, x)\bigg|_{u_0, v_0} (v - v_0). \qquad (6.4b)$$

The linearized equations can now be found. Let $u(x)$ and $v(x)$ be calculated in an iterative scheme; $u_1(x)$ and $v_1(x)$ are values for the first iteration, $u_n(x)$ and $v_n(x)$ values for the nth iteration. Using this symbolism, (6.2a) and (6.2b) can be rewritten as

$$\frac{du_{n+1}(x)}{dx} = \frac{\partial f}{\partial u}(u, v, x)\bigg|_{u_n, v_n} u_{n+1}(x) + \frac{\partial f}{\partial v}(u, v, x)\bigg|_{u_n, v_n} v_{n+1}(x)$$
$$+ f(u_n, v_n, x) - \frac{\partial f}{\partial u}(u, v, x)\bigg|_{u_n, v_n} u_n(x)$$
$$- \frac{\partial f}{\partial v}(u, v, x)\bigg|_{u_n, v_n} v_n(x) + e(x), \qquad (6.5a)$$

$$\frac{dv_{n+1}(x)}{dx} = \frac{\partial g}{\partial u}(u, v, x)\bigg|_{u_n, v_n} u_{n+1}(x) + \frac{\partial g}{\partial v}(u, v, x)\bigg|_{u_n, v_n} v_{n+1}(x)$$
$$+ g(u_n, v_n, x) - \frac{\partial g}{\partial u}(u, v, x)\bigg|_{u_n, v_n} u_n(x)$$
$$- \frac{\partial g}{\partial v}(u, v, x)\bigg|_{u_n, v_n} v_n(x) + f(x). \qquad (6.5b)$$

The linearized differential equations are now in the correct form to apply the recursive equations. $A(x)$, $B(x)$, $C(x)$, $D(x)$, $e(x)$, and $f(x)$, as expressed in (1.1), (1.2), and (6.5) are now functions of $u(x)$ and $v(x)$ as obtained from the previous iteration.

4. SOLUTIONS OF NONLINEAR TWO-POINT BOUNDARY-VALUE PROBLEMS

The linearization process was described in the previous section. Assume that equations of the type given in (6.2a) and (6.2b) are to be solved with the boundary conditions $u(0) = c_1$ and $v(a) = c_2$. The initial approximation $u_0(x) = c_1$ and $v_0(0) = c_2$ is made to start the iterations. It is a fair assumption that the first few iterations will not provide the desired accuracy; thus one proceeds to carry out several iterations before printing the solutions. On the first pass, the transmission matrix and the functions $g(0, x)$ and $h(0, x)$ for $x = a$ are calculated. This gives a second initial guess for $v_1(0)$; this initial value is the only one that must be approximated, since $u(0) = c_1$. Hopefully, $v_1(0)$ is a better approximation than is $v_0(0)$.

The second iteration can now be modified. Since the solutions $u(x)$ and $v(x)$ can be calculated at each point x in the recursive equations, $u_2(x)$ and $v_2(x)$ for $x \neq 0$ can now be found on the second iteration. This means that, aside from the value $v_1(0)$, better values of the transmission matrix and the functions $g(0, x)$ and $h(0, x)$ can be obtained by using $u_2(x)$ and $v_2(x)$. Upon reaching $x = a$ on the second iteration, a new value of $v_2(0)$ is found. This procedure is then continued until the solutions converge.

It should be pointed out that the initial guess, $v_0(0) = c_2$, does not always assure convergence. Certain problems, for which $v(a)$ is large, tend to become unbounded when the recursive equations are used for x increasing in the positive direction. If this happens, use of the recursive equations for x decreasing from $x = a$ to $x = 0$ may be required. The same type of difficulty may be encountered for a backward integration if $u(0)$ is large.

5. EXAMPLE OF A LINEAR TWO-POINT BOUNDARY-VALUE PROBLEM

A simple linear second-order differential equation with a forcing function was selected as an example problem. No attempt was made to choose a well-behaved problem, this problem having been considered by Froberg [1]. An equation which was ill-behaved is mentioned later.

The differential equation

$$\frac{d^2y}{dx^2} + \frac{y(x)}{1 + x^2} = 7x, \qquad (6.6)$$

which has a solution $y(x) = x^3 + x$, was selected as the test problem. The

6 TWO-POINT BOUNDARY-VALUE PROBLEMS

Table 6.1. Solution of Equation (6.6)

$\Delta x = 0.01$, $u(0) = 0$, $v(1) = 4$

x	u(x) (calc.)	u(x) (exact)
0	0.00000000	0.000000000
0.1	0.101000001	0.101000000
0.2	0.208000002	0.208000000
0.3	0.327000002	0.327000000
0.4	0.464000003	0.464000000
0.5	0.625000004	0.625000000
0.6	0.816000004	0.816000000
0.7	1.043000004	1.043000000
0.8	1.312000004	1.312000000
0.9	1.629000004	1.629000000
1.0	2.000000004	2.000000000

two first-order differential equations are

$$\frac{du}{dx} = v(x), \tag{6.7a}$$

$$-\frac{dv}{dx} = \frac{u(x)}{1+x^2} - 7x, \tag{6.7b}$$

where $u(x) = y(x)$ and $v(x) = y'(x)$. The problem was solved over the range $0 \leq x \leq 1$; the boundary conditions are $u(0) = 0$ and $v(1) = 4$. The differential equation in (6.6) is linear, and two passes are required to find the solution. The transmission matrix is calculated on the first pass and the missing value $v(0)$ determined at $x = 1$. This value, which should be correct, is then used to construct the solution on the second pass. The numerical results are given in Table 6.1.

There is an error in the tenth place of the calculated value; it is the result of the inaccuracies in the incremental coefficients. The time to execute the program was 2.4 sec on an SDS Sigma 7 computer.

6. NONLINEAR TWO-POINT BOUNDARY-VALUE PROBLEMS

The solution of a nonlinear two-point boundary-value problem must be found by using an iterative procedure. Since the correct values of $u(x)$ and $v(x)$ are unknown, the two initial-condition vectors (or scalars) cannot be error-free for the first pass. One of the two initial conditions is assumed to be known, the other one is not known.

As mentioned earlier, the calculations are done in an iterative manner. An arbitrary value for the missing initial condition is first selected, the transmission matrix is calculated for each increment, and $u(x)$ and $v(x)$ computed. When the entire medium has been traversed, the final values of the transmission are used to find a new value for the missing initial condition. This procedure is continued until the solutions converge. There is no assurance that the method will converge or that the desired solution will be found. Certain problems, which have been solved using this method, did not converge at all when the initial value was improperly selected. Other problems have been solved which converge to either of two solutions depending on the arbitrary initial condition.

The pendulum problem, discussed earlier, was investigated as a two-point boundary problem. The results obtained from the initial value problem were used for the two-point boundary-values. When the arbitrary initial condition was chosen near the correct value, convergence was noted. If the initial guess was significantly off or if the range on x was too great, convergence was obtained only under certain conditions but not to the desired solution.

The differential equations

$$\frac{du(x)}{dx} = -u^2(x) + v^2(x), \tag{6.8a}$$

$$-\frac{dv(x)}{dx} = -2u(x)v(x), \tag{6.8b}$$

were investigated and found to be well-behaved. Some of the results obtained from the calculations are presented.

The coupled set given in (6.8) was assumed to have the initial conditions $u(0) = 0.50$ and $v(0) = 1.0$. The problem can then be solved for $0 \leq x \leq 1$ to obtain the values $u(1)$ and $v(1)$. The two-point boundary-value problem was then investigated using the values $u(0) = 0.50$ and

$$v(1) = 0.307692307692497.$$

The recursive equations and the quasilinearized differential equations were used with a step size of 0.01. The equations were solved for several initial values of $v(0)$. The estimated value of $v(0)$, the value of $v_n(0)$ calculated for the nth iteration and the values of $u_n(1)$ and $v_n(1)$ for each iteration shown in Table 6.2. Only five iterations have been tabulated, at which point the

6 TWO-POINT BOUNDARY-VALUE PROBLEMS

Table 6.2. Tabulated Values of Nonlinear Two-Point Boundary-Value Problem

(A) Iteration	$u(0)$	$v(0)$	$u(1)$	$v(1)$
1	0.5000	2.0000	0.7600	0.3199
2	0.5000	2.2747	0.7979	0.3063
3	0.5000	2.2501	0.7948	0.3076
4	0.5000	2.2501	0.7948	0.3076
5	0.5000	2.2499	0.7948	0.3076

(B) Iteration	$u(0)$	$v(0)$	$u(1)$	$v(1)$
1	0.5000	1.6000	0.6881	0.3326
2	0.5000	3.4619	0.8946	0.2431
3	0.5000	2.1194	0.7775	0.3143
4	0.5000	2.2546	0.7954	0.3074
5	0.5000	2.2500	0.7948	0.3076

(C) Iteration	$u(0)$	$v(0)$	$u(1)$	$v(1)$
1	0.5000	1.3000	0.6192	0.3299
2	0.5000	0.6830	0.4478	0.2514
3	0.5000	0.9158	0.5143	0.2965
4	0.5000	0.9914	0.5360	0.3066
5	0.5000	0.9998	0.5384	0.3076

(D) Iteration	$u(0)$	$v(0)$	$u(1)$	$v(1)$
1	0.5000	0.1000	0.3362	0.0442
2	0.5000	0.7007	0.4527	0.2556
3	0.5000	0.9230	0.5164	0.2975
4	0.5000	0.9927	0.5364	0.3068
5	0.5000	0.9999	0.5384	0.3076

(E) Iteration	$u(0)$	$v(0)$	$u(1)$	$v(1)$
1	0.5000	−0.2000	0.3449	−0.0873
2	0.5000	0.7373	0.4630	0.2639
3	0.5000	0.9375	0.5206	0.2996
4	0.5000	0.9951	0.5370	0.3071
5	0.5000	0.9999	0.5384	0.3076

(F) Iteration	$u(0)$	$v(0)$	$u(1)$	$v(1)$
1	0.5000	−0.5000	0.4000	−0.2000
2	0.5000	1.0865	0.5627	0.3167
3	0.5000	0.9871	0.5348	0.3061
4	0.5000	0.9997	0.5385	0.3076
5	0.5000	0.9999	0.5384	0.3076

solutions are accurate to four places. Seven iterations were required for fifteen-place accuracy, the sixth and seventh iterations following the quadratic convergence criterion.

The two solutions which were obtained have been plotted in Figure 6.1.

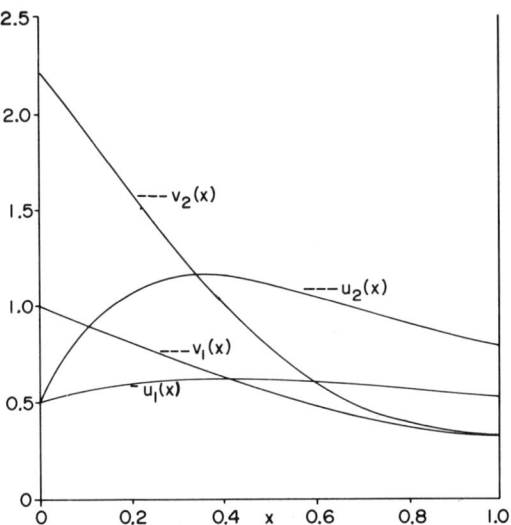

Figure 6.1. Two solutions of nonlinear equations.

The curves labeled $u_1(x)$ and $v_1(x)$ (were calculated with the initial guess of $v(0)$ in the range $-0.5 \leq v(0) \leq 1.3$. These data are given in Table 6.2, parts C, D, and E. The other two curves shown, identified as $u_2(x)$ and $v_2(x)$, are the solutions calculated when the initial guesses for $v(1)$ were 1.6 and 2.0. The constraints placed on $u(0)$ and $v(1)$ require that the two solutions pass through these end points.

The results obtained for the two equations of (6.8) illustrate the behavior of a two-point boundary problem which might be expected. The solutions obtained, provided that convergence occurs, may not be unique solutions. In this case, two different solutions were found. The pendulum problem converged for certain initial values, and at least two solutions were noted.

7. SUMMARY

A study of nonlinear two-point boundary-value problems using the recursive method has been reported in this chapter. The results obtained indicate that much more work is necessary before generalizations can be given for such problems. The linear two-point boundary-value problem is straightforward, and excellent results can be obtained from a two-pass calculation.

The nonlinear problem can be solved if the system of equations is well-behaved, but there is no assurance that the desired solution will be found. The method may converge on only one of the solutions of an entire family. The investigations of nonlinear problems are to be continued and reported on at a later date.

REFERENCE

1. C. E. Froberg, *Introduction to Numerical Analysis*, Addison-Wesley Publishing Company, Reading, Mass., 1965.

CHAPTER 7

Singular Values and Integration in the Complex Plane

The reader has thus far been exposed to the numerical method and sample problems which do not possess troublesome features. Just as in any numerical method, not all problems are so easy to handle as those investigated in earlier sections. This chapter is devoted to discussing some of the difficulties frequently encountered in the numerical method and means of avoiding some of them.

It was noted earlier that the coefficients $A(x)$, $B(x)$, $C(x)$, and $D(x)$ must be bounded at all points where the incremental coefficients are to be calculated. If, for example, any one of these coefficients is unbounded at $x = 0$, the starting values cannot be obtained for the incremental coefficients. The question then naturally arises as how to start solving the recursive equations. Similarly, if any one of the coefficients is unbounded at x_0, and if $n \Delta x$ is selected such that $n \Delta x = x_0$, the coefficient that is unbounded will lead to trouble at that point. This type of coefficients is not so difficult to handle as the unbounded one at $x = 0$, particularly in the case of the initial value problem. The initial value problem requires that $u(0)$ and $v(0)$ be used in the calculations. If $u(x)$ and $v(x)$ are specified at $x \neq 0$, then the problem can be eliminated.

A second problem which frequently arises is that one or more of the transmission matrix coefficients may become singular at some point x_s. If Δx is selected such that $n \Delta x \neq x_s$, the singular point may be avoided by leaping over the singular point. The error introduced by jumping over x_s may be unknown, and such a procedure would be questionable. It is shown here by example that such a procedure may be acceptable, and a method of avoiding the singular point entirely is given. Integration in the complex plane is valid and the singular point can be avoided entirely by suitable check procedures in the program. If any one of the transmission matrices appears to be approaching a singular point, a detour is made into the complex plane. The integration is then continued in the complex plane until the questionable matrix becomes well-behaved. The integration then resumes along the real variable.

I. UNBOUNDED INCREMENTAL COEFFICIENTS

The recursive equations when solved in the positive x direction starting at $x = 0$ depend on evaluation of $A(0)$, $B(0)$, $C(0)$, and $D(0)$. If any coefficients are of infinite value at $x = 0$, the integration procedure cannot be started. There are two rather obvious ways of avoiding this difficulty, one that may introduce a small error and one that should not introduce an error.

If $A(0)$ is infinite at $x = 0$, then this point must be avoided. Assuming that a small error is tolerable, the point $x = 0$ can be approximated by $x = \delta$, where δ is selected so as to approximate zero as far as the computer is concerned; $A(\delta)$ is then finite and integration can proceed. This method should not introduce a large error and may be the most practical one.

Recalling that the incremental coefficients are always evaluated at the last point of integration, it is not difficult to see that if $x = 0$ is approached from the right side, that is, for positive x, the point $x = 0$ is never considered in evaluating $A(x)$. This points up the possibility that integration can proceed from $x = a$, the desired end point, to $x = 0$. Provided that $A(x)$, $B(x)$, $C(x)$, and $D(x)$ are well-behaved for $0 < x \leq a$, no difficulties occur. Such a procedure is equivalent to increasing the medium thickness in the direction of negative x. The differential equations as well as the recursive equations for this case were derived earlier.

If neither of these methods is acceptable, then the Taylor series expansion can be written in a different manner. Consider expanding $R(x, x + \Delta x)$ around the point $x + \Delta x$ rather than x as was done earlier. Then

$$R(x, x + \Delta x) = R(x + \Delta x, x + \Delta x) - R'(x + \Delta x, x + \Delta x)\Delta x$$
$$+ R''(x + \Delta x, x + \Delta x)\frac{\Delta x^2}{2} - \cdots \quad (7.1)$$

and, since $R(x + \Delta x, x + \Delta x) = 0$, we find for second-order incremental coefficients

$$R(x, x + \Delta x) = -A(x + \Delta x)\Delta x + [A'(x + \Delta x) + B(x + \Delta x)A(x + \Delta x)$$
$$+ A(x + \Delta x)C(x + \Delta x)]\frac{\Delta x^2}{2}. \quad (7.2)$$

If $A(x + \Delta x) \neq \infty$, the incremental coefficient is well-behaved. Similarly, the remaining incremental coefficients can be defined by advancing to $x + \Delta x$. By using this technique, any point x which has an ill-behaved coefficient $A(x)$, $B(x)$, $C(x)$, or $D(x)$ can be avoided. The points at which the incremental coefficients cannot be evaluated would have been determined before solving the problem and steps would have been taken in the program to avoid these points.

2. SINGULAR VALUES OF THE MATRIX COEFFICIENTS

To illustrate the presence of singular values in the transmission matrix, consider the simple transport problem (Wing [1]) given below.

$$\frac{du(x)}{dx} = \sigma v(x), \qquad (7.3a)$$

$$-\frac{dv(x)}{dx} = \sigma u(x). \qquad (7.3b)$$

The reflection and transmission coefficients can be derived analytically for this problem; this makes it useful to illustrate the point. Now, from (7.3),

$$\frac{d^2 u(x)}{dx^2} + \sigma^2 u(x) = 0, \qquad (7.4)$$

which has the solution

$$u(x) = c_1 \sin \sigma x + c_2 \cos \sigma x \qquad (7.5a)$$

and

$$v(x) = c_1 \sigma \cos \sigma x - c_2 \sigma \sin \sigma x. \qquad (7.5b)$$

The fundamental matrix can be found from (7.5) and the transmission matrix derived from $\mathbf{Q}(0, x)$. Using the unspecified initial conditions, $\mathbf{Q}(0, x)$ is

$$Q(0, x) = \begin{bmatrix} \cos \sigma x & \dfrac{\sin \sigma x}{\sigma} \\ -\sigma \sin \sigma x & \cos \sigma x \end{bmatrix} \qquad (7.6)$$

and $\mathbf{T}(0, x)$ is obtained from (7.6).

$$\mathbf{T}(0, x) = \begin{bmatrix} \sec \sigma x & \dfrac{\tan \sigma x}{\sigma} \\ \sigma \tan \sigma x & \sec \sigma x \end{bmatrix}. \qquad (7.7)$$

It is apparent that $\mathbf{T}(0, x)$ has singular values of the elements at $\sigma x = \pi/2$. If $n \Delta x \simeq \pi/2\sigma$, the machine indicates an overflow at that point. The actual accuracy in the vicinity of $x = \pi/2\sigma$ is then questionable since the machine is dealing with large numbers.

7 SINGULAR VALUES AND INTEGRATION IN THE COMPLEX PLANE

The fundamental matrix of (7.6) indicates a point which was made earlier: the transmission matrix may have entries that are infinite, whereas the fundamental matrix is well-behaved. This does not imply that the digital machine will not underflow. The fundamental matrix may be somewhat better behaved at $x \simeq \pi/2\sigma$ than is $\mathbf{T}(0, x)$.

One way of avoiding the singular point is to make $x \neq x_s$. A small error may be introduced but, if $n \Delta x$ and $(n + 1) \Delta x$ are sufficiently removed from x_s, the error is small. The problem above was run with $n \Delta x \neq x_s$ with no apparent error. The points x_s and $2x_s$ were crossed and a minimum of eight-place accuracy was obtained. The author was satisfied with the accuracy and considered the amount of error negligible by "leaping" over the singular point. The results of that investigation are given in Table 7.1, where data are shown for $x \leq 2.00$. The data were calculated on an SDS Sigma 7 computer

Table 7.1. Transport Equation

	Step Size: $\Delta x = 0.01$:	$u''(x) + u(x) = 0$
x	$u(x)$ (calc.)	$u(x)$ (corr.)
0.2	0.980066577895174	0.980066577841241
0.4	0.921060994212443	0.921060994002885
0.6	0.825335615364095	0.825335614909678
0.8	0.696706710115772	0.696706709347165
1.0	0.540302306994121	0.540302305868139
1.2	0.362357755972386	0.362357754476673
1.4	0.169967144744443	0.169967142900240
1.6	−0.029199520164346	−0.029199522301288
1.8	−0.227202092351448	—
2.0	−0.416146834118238	−0.416146836547142
	Step Size $\Delta x = 0.001$	
0.2	0.980066577841141	0.980066577841241
0.4	0.921060994002691	0.921060994002885
0.6	0.825335614909397	0.825335614909678
0.8	0.696706609346804	0.696706709347165
1.0	0.540302305867772	0.540302305868139
1.2	0.362357754476301	0.362357754476673
1.4	0.169967142899839	0.169967142900240
1.6	−0.029199522299446	−0.029199522301288
1.8	−0.227202094691237	—
2.0	−0.416146836545364	−0.416146836547142

3. LARGE TRANSMISSION MATRIX COEFFICIENTS AND COMPUTATIONAL ERRORS

The example given in Section 2 was solved for a step size Δx such that $n\,\Delta x \neq x_s$. It is rather interesting to see what happens if $n\,\Delta x = x_s$. Since the computational procedure is not exact, we might expect to obtain large values for the coefficients of the transmission matrix. These values may be of such magnitude that the computer is still handling numbers within its range. No overflow would be indicated and the programmer would not suspect trouble. The example given in Section 6 of Chapter 6 is one in which this occurred, and the computational error was much larger than expected. Consider the equations of (4.41) and the general solutions:

$$\frac{du(x)}{dx} = -3u(x) + v(x) + \exp(2x), \tag{4.41a}$$

$$-\frac{dv(x)}{dx} = u(x) + 5v(x) - \exp(x), \tag{4.41b}$$

$$u(x) = (C_1 + C_2 x)\exp(-4x) + \tfrac{7}{36}\exp(2x) + \tfrac{1}{25}\exp(x), \tag{4.42a}$$

$$v(x) = (C_2 - C_1 - C_2 x)\exp(-4x) - \tfrac{1}{36}\exp(2x) + \tfrac{4}{25}\exp(x). \tag{4.42b}$$

The fundamental matrix is easily determined; it is

$$\mathbf{Q}(0, x) = \begin{bmatrix} (1+x)e^{-4x} & x \\ -x & (1-x)e^{-4x} \end{bmatrix}. \tag{7.8}$$

The fundamental matrix is finite everywhere for finite x. Thus a computational scheme employing the recursive equation of the fundamental matrix would be bounded. On the other hand, the transmission matrix has coefficients that are infinite at $x = 1$. The transmission matrix can be derived from (7.8); it is

$$\mathbf{T}(0, x) = \begin{bmatrix} \dfrac{(1-x^2)e^{-4x} - x^2 e^{4x}}{1-x} & \dfrac{xe^{4x}}{1-x} \\ \dfrac{xe^{4x}}{1-x} & \dfrac{e^{4x}}{1-x} \end{bmatrix}. \tag{7.9}$$

If $x = 1$, all of the coefficients go to an infinite value. With the truncated expansion for the incremental coefficients, the coefficients take on large values but remain finite.

7 SINGULAR VALUES AND INTEGRATION IN THE COMPLEX PLANE

Table 7.2. Computational Error $u(x)$

	Equation (4.41): $\Delta x = 0.001$	
x	$u(x)$ (trans. matrix)	$u(x)$ (fund. matrix)
0.0	1.000000000000	1.000000000000
0.5	0.740966501775	—
1.0	1.571105003356	1.571113724997
1.5	4.089040882859	—
2.0	10.912550895994	10.912550929453
2.5	29.345520662101	—
3.0	79.247925477323	79.247925478242
3.5	214.558845807936	
	$\Delta x = 0.003$	
0.3	0.696103032663	
0.6	0.822385497920	
0.9	1.311196814046	
1.2	2.288755020577	
1.5	4.089041068291	
1.8	7.359731926426	
2.1	13.293904609942	

The important question which must be answered is how the large values degrade the accuracy near $x = 1$ and if the solutions converge for large x. The values obtained in Section 6 of Chapter 6 were computed by use of the fundamental matrix. It was noted in using the transmission matrix that an excessive error appeared at $x = 1$, the trouble arising because of the denominator term of (7.9). Values obtained from the transmission matrix recursive equations are presented in Table 7.2 along with the fundamental matrix values. In this case, Δx was selected to be 0.001, such that $1000 \Delta x = 1$, and also $\Delta x = 0.003$. The first case was picked such that the singular point would be a point for one of the calculations.

4. INTEGRATION IN THE COMPLEX PLANE

The singular points of the transmission matrix can be avoided entirely if the integration procedure allows a detour into the complex plane. The numerical method is valid in the complex plane as well as along the real axis. If the independent variable is $z = x + iy$, then all of the recursive equations, differential equations, and incremental coefficients carry over into the complex plane. Assuming that the singular point occurs at x_s, the detour can be made by leaving the real axis at a point $x_s - k\,\Delta x$, where k is some integer selected to keep the singular coefficient small. Proceed to integrate the equations along y to a point y_1 far enough away from x_s to keep the coefficient small, then turn to integrate along y_1 in the positive x direction. Upon reaching $x_s + k\,\Delta x$, return to the real axis and continue. Such a contour is shown in Figure 7.1. When integrating parallel to the x axis, the recursive

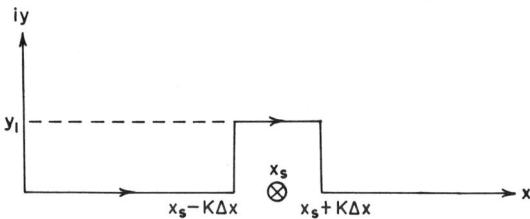

Figure 7.1. Deformed contour.

equations use Δx as the step size, and $i\,\Delta y$ is used when integrating along a path perpendicular to the x axis. The distance y_1 is selected to keep the coefficients of the transmission matrix within a prescribed range.

As an example of the complex plane integration procedure, consider the transport equation given in (7.3) with $\sigma = 1$. The point x_s is located at $\pi/2$, the reflection and transmission coefficients becoming infinite at that point. Equations (7.3a) and (7.3b) in the complex plane are

$$\frac{du(z)}{dz} = v(z), \qquad (7.10a)$$

$$-\frac{dv(z)}{dz} = u(z), \qquad (7.10b)$$

and the transmission matrix is

$$\mathbf{T}(0, z) = \begin{bmatrix} \sec z & \tan z \\ \tan z & \sec z \end{bmatrix}, \qquad (7.11)$$

where $z = x + iy$. Substituting into (7.11) for z and rearranging, we find

$$\mathbf{T}(0, z) = \begin{bmatrix} \dfrac{2\cos x \cosh y + 2i \sin x \sinh y}{\cos 2x + \cosh 2y} & \dfrac{\sin 2x + i \sinh 2y}{\cos 2x + \cosh 2y} \\ \dfrac{\sin 2x + i \sinh 2y}{\cos 2x + \cosh 2y} & \dfrac{2\cos x \cosh y + 2i \sin x \sinh y}{\cos 2x + \cosh 2y} \end{bmatrix}.$$

$$(7.12)$$

7 SINGULAR VALUES AND INTEGRATION IN THE COMPLEX PLANE

Since the transmission matrix is bounded everywhere but at $\cos 2x = -\cosh 2y$, the only point to be avoided is where $\cos 2x = -1$ and $\cosh 2y = 1$ along the x axis. Since $\cosh 2y \geq 1$ for all y, with equality only at $y = 0$ a very small detour suffices.

To test the accuracy of the computational scheme in the complex plane, the example above was solved on a digital computer. The example was solved in a manner slightly different from that shown in Figure 7.1, but this was done for convenience only. The contour of integration was modified to carry out integration along $x = 0$ to a point y_1, then along constant y to $x = \pi$. A value of $y_1 = 0.1$ was picked and a step size of $\Delta x = \pi/100$. Correct values were calculated from (7.10). The results are given in Table 7.3 for the reflection coefficients only.

The fact that the incremental coefficients were written in closed form accounts for the high accuracy. The digits in error are underlined and, as

Table 7.3. Complex Transport Problem

Equation (7.8): $y_1 = 0.1$, $\Delta x = \pi/100$

x		R(z) (calc.)	R(z) (corr.)
$\pi/10$	Real	0.321355023954	0.321355023953
	Imag	0.110074786085	0.110074786100
$2\pi/10$	Real	0.715573052716	0.715573052713
	Imag	0.151484812385	0.151484812407
$3\pi/10$	Real	1.337538619855	1.337538619844
	Imag	0.283153182079	0.283153182118
$4\pi/10$	Real	2.785055281956	2.785055281879
	Imag	0.953974083317	0.953974083430
$5\pi/10$	Real	0.000000000002	0.000000000000
	Imag	10.033311133685	10.033311132254
$6\pi/10$	Real	−2.785055281956	−2.785055281880
	Imag	0.953974083317	0.953974083318
$7\pi/10$	Real	−1.337538619855	−1.337538619844
	Imag	0.283153182079	0.283153182118
$8\pi/10$	Real	−0.715573052716	−0.715573052713
	Imag	0.151484812385	0.151484812407
$9\pi/10$	Real	−0.321355023954	−0.321355023953
	Imag	0.110074786085	0.110074786100
$10\pi/10$	Real	0.000000000000	0.000000000000
	Imag	0.099667994610	0.099667994626

noted, the worst case has nine correct digits. The critical length at which $R(x)$ would become infinite is at $x = \pi/2$; in the complex plane the real part of $R(z)$ vanishes at that point and $T(z)$ showed the same behavior at $x = \pi/2$, as would be expected. The largest value of the imaginary part of $R(z)$ was 10.0 which occurred at $x = \pi/2$. The calculations verify the validity of the computational procedure of avoiding the singular point.

5. THE COMPLEX FRESNEL INTEGRAL

A second example was investigated to verify the applicability of the numerical method in the complex plane. The Fresnel integral is of little interest in the discussion of singularities but, because of the availability of tabulated values, the integral was solved to determine the error.

The complex Fresnel integral is

$$E(z) = \int_0^z \exp\left(i\frac{\pi}{2}t^2\right) dt, \tag{7.13}$$

where $z = x + iy$. Tabulated values of the Fresnel integral have been published by Martz [2]. The values presented in the table are accurate to six decimal places; the calculated values can be checked only to that accuracy.

To solve the integral above, differentiate (7.13) twice with respect to z to set up the two coupled differential equations.

$$\frac{du(x)}{dx} = \exp\left(i\frac{\pi}{2}z^2\right), \tag{7.14a}$$

$$-\frac{dv(x)}{dx} = -i\pi z \exp\left(i\frac{\pi}{2}z^2\right). \tag{7.14b}$$

The equations are now in the correct form if we let $\exp(i/2\pi z^2) = v(x)$. The fourth-order incremental coefficients can be derived from (7.14) and (4.64).

$$\mathbf{T}(z, z + \Delta z) = \begin{bmatrix} 1 & 0 \\ 0 & 1 \end{bmatrix} + \begin{bmatrix} 0 & 1 \\ 0 & -i\pi z \end{bmatrix} \Delta z + \begin{bmatrix} 0 & -i\pi z \\ 0 & -i\pi - \pi^2 z^2 \end{bmatrix}\frac{\Delta z^2}{2}$$
$$+ \begin{bmatrix} 0 & -2i\pi - \pi^2 z^2 \\ 0 & i\pi^3 z^3 - 3\pi^2 z^2 \end{bmatrix}\frac{\Delta z^3}{6} + \begin{bmatrix} 0 & -5\pi^2 z + i\pi^3 z^3 \\ 0 & -3\pi^2 + 6i\pi^3 z^2 + \pi^4 z^4 \end{bmatrix}\frac{\Delta z^4}{24}. \tag{7.15}$$

The complex Fresnel integral was solved for a limited number of values to prove the accuracy. The equation was first solved along $x = 0$, starting at

7 SINGULAR VALUES AND INTEGRATION IN THE COMPLEX PLANE

$y = 0$, out to $y = 1.0$. The calculated values along $(0, y)$ were then used as initial conditions to integrate along x for a constant y. The initial values at $(0, 0)$ are

$$u(0, 0) = 0 + i0,$$
$$v(0, 0) = 1 + i0.$$

The fundamental matrix at the point $(0, y)$ is needed to start the integration along a constant y; in this case y was picked to be 0.2. The fundamental matrix at $(0, 0.2)$ is

$$\mathbf{Q}(0, 0.2) = \begin{bmatrix} 1.0 & 0.004187609 + i0.1999210575 \\ 0 & 0.9980267284 - i0.0627905195 \end{bmatrix}.$$

This initial condition was then used to proceed along x to find $\mathbf{Q}(x, 0.2)$. The values of $u(x, y)$ and $v(x, y)$ are then calculated from

$$\begin{bmatrix} u(x, y) \\ v(x, y) \end{bmatrix} = [\mathbf{Q}(x, 0.2)][\mathbf{Q}(0, 0.2)] \begin{bmatrix} u(0, 0) \\ v(0, 0) \end{bmatrix}. \tag{7.16}$$

Values of $E(x, y)$ are tabulated in Table 7.4 along with the values of

Table 7.4. Complex Fresnel Integral

(x, y)		$E(x, y)$ (calc.)	$E(x, y)$ (Martz)
(0.2, 0.2)	Real	0.19192902	0.19193
	Imag	0.19192902	0.19193
(0.4, 0.2)	Real	0.35687996	0.35688
	Imag	0.20545182	0.20545
(0.6, 0.2)	Real	0.49451728	0.49452
	Imag	0.25287213	0.25287
(0.8, 0.2)	Real	0.59170952	0.59171
	Imag	0.33598060	0.33598
(1.0, 0.2)	Real	0.63144500	0.63144
	Imag	0.44089333	0.44089

Martz. The values calculated from the recursive equations agree with the values published by Martz. The total time to run the program above, including compiling, execution, and printing, was 90 sec on an IBM 360 Model 44 computer. Values of $E(z)$ were printed out at other points but are not given here.

6. SUMMARY

Numerical integration in the complex plane has been described in this chapter. It is sometimes necessary to avoid singular points, and the detour into the complex plane appears to be very useful for bypassing the singular points. The entire question of how far this method can be pushed is still open and much work remains to be done on investigating the procedure in the complex plane.

The writer carried out a limited investigation of contour integration in the complex plane with no conclusive results. Many functions can be evaluated in terms of a contour integration about a singular point. In some cases the numerical integration gave good results. In a few cases the contour integral could not be evaluated to the desired accuracy. Whether the method failed or the function was poorly behaved is not known, since there was not time to complete the investigations.

Evaluation of certain functions in the complex plane, such as the error integral, was successful. There are other functions which could be calculated in this manner; the application is left to the interested reader. It is not the intention of this work to present tabulated values of functions but to present the method.

REFERENCES

1. G. M. Wing, *Introduction to Transport Theory*, John Wiley and Sons, New York, 1962.
2. C. W. Martz, Tables of the Complex Fresnel Integral, National Aeronautics and Space Administration, Rept. SP-3010, Washington, D.C.

CHAPTER 8

The Recursive Procedure Compared to Other Numerical Methods

The properties which a numerical method should possess are: accuracy, simplicity, and adaptability. The numerical method should provide accurate answers to the problem without carrying out an excessive number of arithmetic operations. All digital machines are limited in the number of bits that can be used, and the accuracy of the add, subtract, multiply, and divide operations depends on the word length. It is therefore advisable to keep the number of arithmetic operations within some reasonable limit.

The property of simplicity may not seem to be important to an experienced programmer once he has mastered a method. To the part-time programmer, it is quite important since the amount of time that he has available to spend on learning a technique is limited. A researcher analyzing physical systems may not need a numerical method in his everyday assignment, yet there are times when he must solve a particular set of differential equations. It is to his advantage if the method is simple so that he can master the technique in a short time. The same argument is valid for students learning to program. Much of the student's time is spent on organizing the program logic, not in learning how numerical methods are constructed.

The adaptability property is perhaps the most important of all of the properties which a numerical method should have. If each problem to be solved calls for a new method, much time will be lost in mastering each of the required techniques. A full-time professional programming staff may have the manpower to provide a specialist for each numerical method, but many organizations cannot afford this staff coverage.

The one remaining property, which encompasses the three properties above, is that of computation time for any numerical method. The amount of time spent to carry out the required computations can be quite important. The operational cost of a large computer system is high. Since the operational cost is almost fixed for a modern computer center, the economy depends on the number of jobs completed during a work period. A 10 percent saving in computational time for the same accuracy may lengthen the working life of a computer by several months.

This chapter is devoted to describing two of the better-known numerical methods, Runge-Kutta and Adams-Moulton, and comparing some of the results obtained using these two methods. A comparison is given for two problems solved using the Runge-Kutta, the Adams-Moulton, and the recursive scheme. The reader should realize that an exhaustive comparison requires a great amount of time and effort. The material to be presented can only consider *some* of the characteristics of each method.

I. RUNGE-KUTTA METHOD

The numerical method of Runge-Kutta has been in use for many years. It is a commonly used method and there are different versions of Runge-Kutta. The one to be described here is the fourth-order method. This order compares favorably with the invariant imbedding technique in accuracy and in the number of calculations per step Δx for a coupled set of equations.

The coupled differential equations considered in this book can be written at

$$\frac{du(x)}{dx} = f(u, v, x) = B(x)u(x) + A(x)v(x), \tag{8.1a}$$

$$\frac{dv(x)}{dx} = g(u, v, x) = -D(x)u(x) - C(x)v(x). \tag{8.1b}$$

It is assumed that both equations must be solved to describe the system.

The Runge-Kutta equations for the two coupled equations are

$$u_{n+1} = u_n + \tfrac{1}{6}(k_1 + 2k_2 + 2k_3 + k_4), \tag{8.2a}$$

$$v_{n+1} = v_n + \tfrac{1}{6}(l_1 + 2l_2 + 2l_3 + l_4), \tag{8.2b}$$

where

$$k_1 = f(u_n, v_n, x_n)\,\Delta x, \tag{8.3a}$$

$$k_2 = f\left(u_n + \frac{k_1}{2}, v_n + \frac{l_1}{2}, x_n + \frac{\Delta x}{2}\right)\Delta x, \tag{8.3b}$$

$$k_3 = f\left(u_n + \frac{k_2}{2}, v_n + \frac{l_2}{2}, x_n + \frac{\Delta x}{2}\right)\Delta x, \tag{8.3c}$$

$$k_4 = f(u_n + k_3, v_n + l_3, x_n + \Delta x)\,\Delta x, \tag{8.3d}$$

and a similar set of equations describe l_1, l_2, l_3, and l_4, where l_i is defined in terms of $g(u, v, x)$.

8 THE RECURSIVE PROCEDURE COMPARED TO OTHER NUMERICAL METHODS

Ten equations are required to compute u_{n+1} and v_{n+1} when the past values of u_n and v_n are known. This is the same number of equations required when the recursive equations are solved. The major difference in the two methods is not indicated by the number of equations to be solved. The incremental coefficients which are required in the recursive equations method can be worked out by hand or they can be formed in the program. In the event that the incremental coefficient equations are not set up within the program, considerable hand labor is involved. If the equations are set up within the program, a larger number of multiplications and other arithmetic operations is required as compared to the Runge-Kutta program. Runge-Kutta has an advantage in this respect.

If $A(x)$, $B(x)$, $C(x)$, and $D(x)$ are not spatially varying, but constants, the invariant imbedding method has a slight advantage. The incremental coefficients are now constants for all x and must be evaluated only once in a program. This evaluation can be done in the initial phase of the program. The number of equations for the Runge-Kutta method remains fixed, since the functionals $f(u, v, x)$ and $g(u, v, x)$ must still be evaluated at each x. The number of equations in the recursive method for each value of x then becomes six.

One of the differences in the two methods is seen when (8.2) and the equations

$$u(x) = \tau(0, x)u(0) + R(0, x)v(x), \tag{3.15a}$$

$$v(0) = \mathcal{R}(0, x)u(0) + T(0, x)v(x), \tag{3.15b}$$

are compared. The transmission matrix is independent of $u(x)$ and $v(x)$ when the coupled equations are linear. The Runge-Kutta method requires that all values of $u(x)$ and $v(x)$ be generated in the integration. The recursive scheme requires that only those values of $u(x)$ and $v(x)$ are to be calculated that are needed and, furthermore, an entire family of solutions can be generated when $u(0)$ and $v(0)$ are varied. The Runge-Kutta method does not permit this generation of a family of solutions since changing $u(0)$ and $v(0)$ requires restarting the integration.

Two-point boundary problems can be solved using the method of invariant imbedding but cannot be solved directly by the Runge-Kutta method. The auxiliary solutions of $R(0, x)$, $T(0, x)$, $\mathcal{R}(0, x)$, and $\tau(0, x)$ for a linear problem do not have a functional dependency on $u(x)$ and $v(x)$. This makes it possible to calculate the transmission matrix and then to determine the missing boundary conditions. The Runge-Kutta numerical method requires that $k_i(u, v, x)$ and $l_i(u, v, x)$ be generated. This functional dependency makes it impossible to solve boundary-value problems directly. This gives the recursive method a clear advantage in adaptability.

126 COUPLED MODES IN PLASMAS, ELASTIC MEDIA, PARAMETRIC AMPLIFIERS

Both Runge-Kutta and invariant imbedding have self-starting features for initial value problems. There is little difference between them in this feature other than the clear difference between the functional dependency of $k_i(u, v, x)$ and $l_i(u, v, x)$ and the transmission matrix. The functions $k_i(u, v, x)$ and $l_i(u, v, x)$ can be calculated if u_0, v_0 are given for x_0, where x_0 is the starting point of the calculations.

2 AN EXAMPLE PROBLEM FOR COMPARING THE RUNGE-KUTTA AND THE INVARIANT IMBEDDING METHODS

To illustrate the difference between Runge-Kutta and invariant imbedding, a sample problem was selected and solved numerically on a Sigma 7 computer.

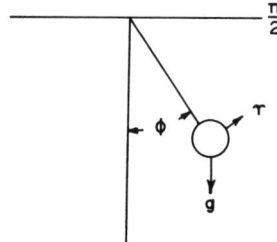

Figure 8.1. Simple pendulum with constant torque.

Consider the pendulum problem illustrated in Figure 8.1. The equation of motion for the pendulum is

$$ml^2 \frac{d^2\phi}{dt^2} + b\phi + mgl \sin \phi = \Upsilon \tag{8.4}$$

where m is the mass, l is the length of the arm, b is the viscous damping coefficient, g is the acceleration of gravity and, Υ is the applied torque. The equation can be modified by letting

$$x = t\sqrt{\frac{g}{l}}, \tag{8.5a}$$

$$\frac{d\phi}{dt} = \frac{d\phi}{dx}\sqrt{\frac{g}{l}}, \tag{8.5b}$$

and

$$\frac{d^2\phi}{dt^2} = \frac{g}{l}\frac{d^2\phi}{dx^2}. \tag{8.5c}$$

The equation of motion then becomes

$$\frac{d^2\phi}{dx^2} + k_1 \frac{d\phi}{dx} + \sin\phi = k_2, \qquad (8.6)$$

where

$$k_1 = \frac{b}{ml^2}\sqrt{\frac{l}{g}}, \qquad k_2 = \frac{\Upsilon}{mlg}.$$

The second-order equation of (8.6) can be solved as it stands by the Runge-Kutta method. The Runge-Kutta equations for (8.6) and the equation $u'' = f(u, u', x)$ are given by

$$\phi_{n+1} = \phi_n + [\phi_n' + \tfrac{1}{6}(k_1 + k_2 + k_3)]\,\Delta x, \qquad (8.7a)$$

$$\phi_{n+1}' = \phi_n' + \tfrac{1}{6}(k_1 + 2k_2 + 2k_3 + k_4), \qquad (8.7b)$$

$$k_1 = f(\phi_n, \phi_n', x_n)\,\Delta x, \qquad (8.7c)$$

$$k_2 = f\left(\phi_n + \frac{\Delta x}{2}\phi_n' + \frac{\Delta x}{8}k_1,\, \phi_n' + \frac{k_1}{2},\, x_n + \frac{\Delta x}{2}\right)\Delta x, \qquad (8.7d)$$

$$k_3 = f\left(\phi_n + \frac{\Delta x}{2}\phi_n' + \frac{\Delta x}{8}k_2,\, \phi_n' + \frac{k_2}{2},\, x_n + \frac{\Delta x}{2}\right)\Delta x, \qquad (8.7e)$$

$$k_4 = f\left(\phi_n + \Delta x\phi_n' + \frac{\Delta x}{2}k_3,\, \phi_n' + k_3,\, x_n + \Delta x\right)\Delta x. \qquad (8.7f)$$

This formulation should be as efficient as solving two first-order differential equations by means of Runge-Kutta. This set of equations was solved to give a comparison between the Runge-Kutta and invariant imbedding methods. Both the accuracy and the computational time were considered in the comparison.

To solve the second-order equation by invariant imbedding, two first-order equations had to be generated. The coupled differential equations below were used in the invariant imbedding approach.

$$\frac{d\phi}{dx} = \theta, \qquad (8.8a)$$

$$-\frac{d\theta}{dx} = k_1\theta + \sin\phi - k_2. \qquad (8.8b)$$

If θ and ϕ are considered as the two variables, two options are available when the set is solved. If k_2 is to be assigned as the forcing function, then (8.8b) must be rearranged. Multiply and divide $\sin\phi$ by ϕ, thus obtaining

$$-\frac{d\theta}{dx} = k_1\theta + \left(\frac{\sin\phi}{\phi}\right)\phi - k_2, \qquad (8.9)$$

where k_1 can be considered to be the coefficient C and $\sin \phi/\phi$ the coefficient $D(\psi)$. The other possibility is to assign $\sin \phi$ and k_2 to the forcing function. The latter approach was taken in solving the equations of (8.8).

The results of the numerical experiment are shown in Table 8.1. A step size of $\Delta x = 0.01$ was used and values of $k_1 = 0.3$ and $k_2 = 0.5$ were selected. Two initial conditions are required; values of $\phi(0) = 2.63$ and $\dot{\phi}(0) = 0$ were picked. The solutions were also obtained by means of the Runge-Kutta method for a step size of 0.001 to determine a more accurate solution. Double precision was used in both runs.

The two methods gave results which are identical to eleven decimal places. There is a small difference in the last place of the calculated results. The exact

Table 8.1. Pendulum Problem

$\Delta x = 0.01, \quad k_1 = 0.3, \quad k_2 = 0.5$

Values of $\phi(\tau)$

τ	Runge-Kutta	Inv. Imb.
0	2.63000000000	2.63000000000
0.5	2.63126382645	2.63126382645
1.0	2.63507877588	2.63507877588
1.5	2.64187762723	2.64187762723
2.0	2.65265368805	2.65265368806
2.5	2.66909744655	2.66909744657
3.0	2.69387063142	2.69387063145
3.5	2.73107013710	2.73107013715
4.0	2.78698239081	2.78698239089

Values of $\dot{\phi}(\tau)$

0	0.00000000000	0.00000000000
0.5	0.00502219721	0.00502219721
1.0	0.01038121573	0.01038121573
1.5	0.01714082270	0.01714082270
2.0	0.02651802037	0.0265180203
2.5	0.04012924967	0.04012924968
3.0	0.06030478129	0.06030478129
3.5	0.09054937021	0.09054937023
4.0	0.13626811122	0.13626811126

Computation time for $0 \leq \tau \leq 4.9$: Runge-Kutta, 1.92 sec; invariant imbedding, 1.56 sec solution for this problem is not available, and the Runge-Kutta calculations with a step size of 0.001 were used as the exact answers. Using these values as correct, the invariant imbedding has the larger error. However, there is no assurance that the Runge-Kutta calculations for $\Delta x = 0.001$ are correct. The conclusion reached is therefore only an observation.

The computation times listed at the bottom of Table 8.1 are given for the problem run on the same computer and during the same day. Invariant imbedding has a slight advantage. This is in opposition to some of the remarks made earlier. Why the invariant imbedding method runs faster in this problem is unknown.

3. THE ADAMS-MOULTON METHOD

The second and last type of numerical method considered in this limited study was the Adams modified method. This method is a predictor-corrector numerical scheme which is not self-starting. Some other technique must be used to carry the calculations through the first few steps before the Adams method can be applied. Although this is not a serious disadvantage, it makes the program somewhat more complex. A Taylor series or the Runge-Kutta method can be used as a starting device.

The predictor-corrector method is to estimate values of u_{n+1} and v_{n+1} at $(n + 1) \Delta x$ and then to calculate u'_{n+1} and v'_{n+1} from the equations

$$u'_{n+1} = f(u_{n+1}, v_{n+1}, x), \tag{8.10a}$$

$$v'_{n+1} = g(u_{n+1}, v_{n+1}, x), \tag{8.10b}$$

where $x = (n + 1) \Delta x$. Higher-order differences, i.e., $\nabla^k u'_{n+1}$ and $\nabla^k v'_{n+1}$, are formed from u'_{n+1} and v'_{n+1}. The new estimates for u_{n+1} and v_{n+1} are then found. This cycle is repeated until the desired accuracy is obtained.

There are several ways of obtaining the first estimates of u_{n+1} and v_{n+1}. Assume that u_n and v_n are known and that the higher-order differences can be calculated. The first estimate to u_{n+1} may then be

$$u_{n+1} \simeq u_n + (1 + \tfrac{1}{2}\nabla + \tfrac{5}{12}\nabla^2 + \cdots) \Delta x, \tag{8.11}$$

where ∇ indicates the backward difference.

$$\nabla u(x) = u(x) - u(x - \Delta x), \tag{8.12a}$$

$$\nabla^{r+1} u(x) = \nabla^r u(x) - \nabla^r u(x - \Delta x). \tag{8.12b}$$

The estimate obtained in (8.11) is then used in (8.10a) with a similar calculation to be made for v'_{n+1}. The values of u'_{n+1} and v'_{n+1} are then available. A second equation,

$$u_{n+1} \simeq u_n + (1 - \tfrac{1}{2}\nabla - \tfrac{1}{12}\nabla^2 - \tfrac{1}{24}\nabla^3 - \cdots)u'_{n+1}\,\Delta x, \qquad (8.13)$$

is then solved for the next estimate of u_{n+1} with a similar equation for v_{n+1}.

The new estimates can then be used to calculate better estimates. This procedure is continued until some set error criterion is achieved. If u_{n+1} is the next estimate such that $u_{n+1} - u_n < \varepsilon$, where ε is a set error, the procedure is terminated.

The Adams method does not require a large number of equations to be programmed since much of the work can be done in a loop. This does not imply that there are fewer arithmetic operations. The number of calculations is set by the error criterion and by how rapidly the estimates converge to satisfy the error criterion.

Initial values of $u(x)$ and $v(x)$ are required before the solution can be started. The method is unsuitable for two-point boundary-value problems just as is the Runge-Kutta method. This gives the invariant imbedding approach an advantage over the Adams method for two-point boundary value problems.

The error of the modified Adams method is approximately $0(\Delta x^5)$ or smaller depending on the truncation of (8.11) and (8.13). The series should be calculated to the desired accuracy by carrying along the differences necessary to obtain the desired accuracy.

4. NUMERICAL COMPARISON OF SEVERAL EQUATIONS

To obtain some data on the accuracy of the Adams method, several equations were solved by that method. The same equations were solved by Runge-Kutta and by the invariant imbedding method. Computational times are not given, since the problems were run on different machines and the Adams program used was a general program for a larger number of equations than considered in the invariant imbedding scheme.

The two equations studied were the first- and second-order differential equations below.

$$\frac{dy}{dx} = -2xy^2, \qquad (8.14)$$

$$\frac{d^2y}{dx^2} + 8\frac{dy}{dx} - 20y = 0. \qquad (8.15)$$

The first of these two equations was discussed in Chapter 5. The second

8 THE RECURSIVE PROCEDURE COMPARED TO OTHER NUMERICAL METHODS

equation was studied in detail because of the behavior of the two solutions. The roots of the characteristic equation

$$\lambda^2 + 8\lambda - 20\lambda = 0 \tag{8.16}$$

are $\lambda = 2$ and $\lambda = -10$. The general solution of (8.15) is then given by

$$y(x) = c_1 e^{2x} + c_2 e^{-10x}. \tag{8.17}$$

If $c_2 = 0$, only the term $\exp(2x)$ is present in the solution. For $c_1 = 0$, the

Table 8.2. Differential Equation with Constant Coefficient

x	Exact	Runge-Kutta	Adams	Inv. Imb.
(A) $\Delta x = 0.01$				
1	$7.3890560989E + 00$	$7.3890560796E + 00$	$7.3890561522E + 00$	$7.3890575872E + 00$
2	$5.4598150033E + 01$	$5.4598149747E + 01$	$5.4598150833E + 01$	$5.4598177680E + 01$
3	$4.0342879349E + 02$	$4.0342880241E + 02$	$4.0342880241E + 02$	$4.0342912080E + 02$
4	$2.9809579870E + 03$	$2.9809579558E + 03$	$2.9809580751E + 03$	$2.9809613146E + 03$
5	$2.2026465794E + 04$	$2.2026465506E + 04$	$2.2026466609E + 04$	$2.2026497099E + 04$
(B) $\Delta x = 0.001$				
1	$7.3890560989E + 00$	$7.3890560989E + 00$	$7.3890560989E + 00$	$7.3890560990E + 00$
2	$5.4590150033E + 01$	$5.4598150033E + 01$	$5.4598150033E + 01$	$5.4598150035E + 01$
3	$4.0342879349E + 02$	$4.0342879349E + 02$	$4.0342879349E + 02$	$4.0342879352E + 02$
4	$2.9809579870E + 03$	$2.9809579870E + 03$	$2.9809579870E + 03$	$2.9809579873E + 03$
5	$2.2026465794E + 04$	$2.2026465795E + 04$	$2.2026465794E + 04$	$2.2026465797E + 04$

solution is given by the second term on the right side of (8.17). If an attempt is made to calculate the $\exp(-10x)$ solution, the $\exp(2x)$ solution introduces computational instabilities. The invariant imbedding, Runge-Kutta, and Adams methods were all tried on this problem and all had the same difficulty. In no case could the solution $\exp(-10x)$ be followed beyond $x = 3$.

The solution $\exp(2x)$ could be calculated, and the results for that work are reported here. The exact solution for $c_1 = 1$ and $c_2 = 0$, the calculated values via Runge-Kutta, Adams, and invariant imbedding are given in Table 8.2. The invariant imbedding method had the largest error of the three methods for this problem. The larger error is due to the slow convergence for the Taylor series which is used to obtain the incremental coefficients. A higher-order Taylor series would give better results, but a series representation beyond the fourth order is not desirable for the incremental coefficients.

The solutions were obtained for two step sizes, 0.01 and 0.001. Double precision was used in the runs, all of which were carried out on an IBM 360 Model 44 computer.

Table 8.3. $(1 + x^2)^{-1}$ **Solution of Equation (8.14)**

		$\Delta x = 0.01$		
x	Exact	Runge-Kutta	Adams	Inv. Imb.
1	0.50000000000	0.50000000007	0.49999999864	0.50000000000
2	0.20000000000	0.20000000006	0.20000000001	0.20000000000
3	0.10000000000	0.10000000002	0.09999999996	0.10000000000
4	0.05882352941	0.05882352942	0.05882352939	0.05882352941
5	0.03846153846	0.03846153846	0.03846153845	0.03846153846

The nonlinear equation of (8.14) was solved by the three methods, and in this case invariant imbedding was more accurate. The Taylor series converges quite rapidly and very little error is noted in the calculations. Fourth-order incremental coefficients were used for the invariant imbedding method; the results for $\Delta x = 0.01$ are given in Table 8.3. Twelve-place agreement was obtained for the three methods when $\Delta x = 0.001$; the results are not given since there was excellent agreement between the three methods.

5. SUMMARY

This chapter has discussed some of the properties of Runge-Kutta, Adams-Moulton, (see [1]–[3]) and invariant imbedding numerical methods. The results obtained were rather inconclusive because of the examples selected. In one case, the Runge-Kutta and Adams methods were clearly more accurate. The other case revealed that invariant imbedding gave the most accurate answers. The only conclusion reached in the work is that the accuracy depends on the accuracy of the incremental coefficients. This indicates that, if the Taylor series for the incremental coefficients of the problem is slowly converging, methods other than invariant imbedding will probably give better results.

The invariant imbedding method can be used for two-point boundary problems. The other methods do not appear to be suitable for this type of problem. The invariant imbedding method cannot be justified solely on that basis since the majority of differential equations encountered are not of this type.

A rather interesting fact is that the invariant imbedding method was somewhat faster than the Runge-Kutta method. Why this is true is not clear, since it appears that fewer arithmetic operations are required in Runge-Kutta. This observation needs to be studied further.

REFERENCES

1. C. E. Froberg, *Introduction to Numerical Analysis*, Addison-Wesley Publishing Company, Reading, Mass, 1965.
2. F. B. Hildebrand, *Introduction to Numerical Analysis*, McGraw-Hill Book Company, New York, 1965.
3. *N.B.S. Handbook of Mathematical Functions*, National Bureau of Standards, Applied Math. Series No. 55, Washington, D.C., 1964.

CHAPTER 9

Transmission Matrices for Circuits and Discrete Media

The solution of coupled differential equations was considered in the first part of this book. The "wave" medium was considered to be continuous and the incremental coefficients found by a Taylor series expansion. There are many problems which must be defined in a discrete sense. Attention is now turned to them. If the medium is discrete, that is, if short sections of transmission line with unequal impedances are cascaded, the discrete versions of the recursive equations must be used. The incremental coefficients cannot be calculated by a Taylor series since the differential equation representation is missing. The transmission line then consists of short sections with a finite change of impedance at each discontinuity. This discontinuity property was discussed in Chapter 3 for the single-mode problem.

Many problems, aside from the transmission line problem, must be defined in terms of discrete electrical elements. The parametric amplifier is a good example of a coupled mode problem in which lumped electrical elements are present. If the scattering matrix can be found for the electrical elements [1], the cascading process can be carried out. Linear time-invariant elements are not difficult to handle; nonlinear elements, such as voltage variable capacitors, are more difficult to treat.

Some of the methods of obtaining the transmission matrix for a lumped electrical element or discrete section of line are presented here. The treatment is not extensive; only some of the obvious problems are considered. The reader may find other ways of deriving the desired coefficients. In most cases, the type of problem determines the method of obtaining the coefficients.

I. TIME-INVARIANT COMPONENTS AND THE SCATTERING MATRIX

It has been shown that the scattering and transmission matrix are related. If the scattering matrix can be found, the transmission matrix is known. To

9 TRANSMISSION MATRICES FOR CIRCUITS AND DISCRETE MEDIA

illustrate some of the fundamental ideas, consider the two-port network shown in Figure 9.1. The waves a_1, a_2, b_1, and b_2 are incident and reflected waves, respectively. The reflected waves are given by the product of the scattering matrix and the incident wave vector.

$$\begin{bmatrix} b_1 \\ b_2 \end{bmatrix} = \begin{bmatrix} s_{11} & s_{12} \\ s_{21} & s_{22} \end{bmatrix} \begin{bmatrix} a_1 \\ a_2 \end{bmatrix}. \tag{9.1}$$

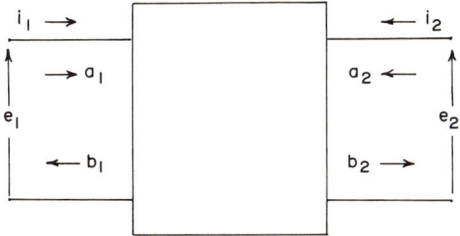

Figure 9.1. Two-port network.

The normalized waves are considered; that is,

$$a_1 = \frac{1}{2}\left[\frac{e_1}{\sqrt{r_{01}}} + i_1\sqrt{r_{01}}\right], \tag{9.2a}$$

$$b_1 = \frac{1}{2}\left[\frac{e_1}{\sqrt{r_{01}}} - i_1\sqrt{r_{01}}\right], \tag{9.2b}$$

$$a_2 = \frac{1}{2}\left[\frac{e_2}{\sqrt{r_{02}}} + i_2\sqrt{r_{02}}\right], \tag{9.2c}$$

$$b_2 = \frac{1}{2}\left[\frac{e_2}{\sqrt{r_{02}}} - i_2\sqrt{r_{02}}\right], \tag{9.2d}$$

where r_{01} and r_{02} are normalizing constants, usually chosen to obtain the simplest form for the power relations. The normalized scattering matrices are generally discussed in books and articles on scattering matrices. Since the recursive equations are not normalized, the nonnormalized scattering matrices will be found later.

Assume that the line to the left of the two-port has a normalizing impedance r_0 and an impedance r_{02} on the right side. The normalized scattering matrix is then given by

$$\mathbf{S} = (Z + I)^{-1}(Z - I), \tag{9.3}$$

where Z is the normalized impedance matrix of the two-port network,

$$Z = R_0^{-1/2} \underline{Z} R_0^{-1/2}, \tag{9.4}$$

$\underline{Z}$ is the nonnormalized impedance of the network, and $R_0^{-1/2}$ is given by the diagonalized matrix in (9.5).

$$R_0^{-1/2} = \begin{bmatrix} \dfrac{1}{\sqrt{r_{01}}} & 0 \\ 0 & \dfrac{1}{\sqrt{r_{02}}} \end{bmatrix}. \tag{9.5}$$

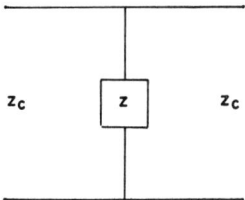

Figure 9.2. Shunt impedance.

Equation (9.3) can also be written as

$$S = [I + Y]^{-1}[I - Y], \tag{9.6}$$

where Y is the normalized admittance matrix for the network, i.e., $Y = Z^{-1}$.

The nonnormalized scattering matrix can be found once the normalized matrix is known. The nonnormalized scattering matrix is given by

$$\underline{S} = R_0^{1/2} S R_0^{-1/2}, \tag{9.7}$$

and the nonnormalized impedance-scattering relation is

$$\underline{S} = (\underline{Z} - R_0)(\underline{Z} + R_0)^{-1}. \tag{9.8}$$

Consider now the two-port network with a shunt element z connected across a transmission line with a characteristic impedance z_c. The matrix R_0 is then

$$R_0 = \begin{bmatrix} z_c & 0 \\ 0 & z_c \end{bmatrix}; \tag{9.9}$$

the two-port network is given in Figure 9.2.

9 TRANSMISSION MATRICES FOR CIRCUITS AND DISCRETE MEDIA

The impedance matrix of the network is

$$\underline{Z} = \begin{bmatrix} z & z \\ z & z \end{bmatrix} \quad (9.10)$$

and the nonnormalized scattering matrix is

$$\underline{S} = \begin{bmatrix} z - z_c & z \\ z & z - z_c \end{bmatrix} \begin{bmatrix} z + z_c & z \\ z & z + z_c \end{bmatrix}^{-1} \quad (9.11)$$

or

$$\underline{S} = \frac{1}{2z + z_c} \begin{bmatrix} -z_c & z \\ z & -z_c \end{bmatrix}. \quad (9.12)$$

Equation (9.12) is the nonnormalized scattering matrix for the circuit shown in Figure 9.2.

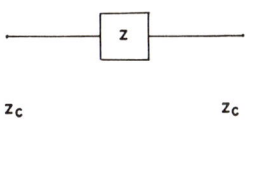

Figure 9.3. Series impedance.

A series element in a transmission line, as shown in Figure 9.3, does not have an impedance matrix. The network does, however, possess an admittance matrix and a scattering matrix. To find the nonnormalized scattering matrix, the admittance matrix must be used. Equation (9.8) can be written as

$$\underline{S} = (I - R_0 Y)(I + R_0 Y)^{-1}, \quad (9.13)$$

where Y for the network of Figure 9.3 is given by

$$Y = \begin{bmatrix} y & -y \\ -y & y \end{bmatrix}. \quad (9.14)$$

The matrix R_0 is defined as in (9.9) with the characteristic impedance z_c on both sides of the network. Using (9.9) and (9.14), one finds the scattering matrix:

$$\underline{S} = \frac{1}{z + 2z_c} \begin{bmatrix} z & 2z_c \\ 2z_c & z \end{bmatrix}. \quad (9.15)$$

Networks with series and shunt elements can now be analyzed by using the discrete recursive equations. Assume that the network given in Figure 9.4 is to be analyzed and the transmission matrix found. It was shown earlier that the transmission matrix can be found from the scattering matrix.

$$\mathbf{T} = \begin{bmatrix} 0 & 1 \\ 1 & 0 \end{bmatrix} \begin{bmatrix} S_{11} & S_{12} \\ S_{21} & S_{22} \end{bmatrix}. \tag{9.16}$$

The composite **T** matrix for the two networks in cascade is then found by

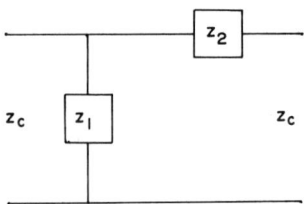

Figure 9.4. Shunt-series two-port network.

using the recursive relations; **S**(1) is defined by (9.12), and **S**(2) by (9.15). The matrix **T**(1) can be found by using the transformation of (9.16):

$$\mathbf{T}(1) = \frac{1}{z_1 + 2z_c} \begin{bmatrix} 2z_c & z_1 \\ z_1 & 2z_c \end{bmatrix}. \tag{9.17}$$

The transmission matrix for the shunt element in Figure 9.4 is

$$\mathbf{T}(2) = \frac{1}{2z_2 + z_c} \begin{bmatrix} z_2 & -z_c \\ -z_c & z_2 \end{bmatrix}. \tag{9.18}$$

The composite transmission matrix **T**(1, 2) is then found from (3.7), where the elements are joined at the common node.

$$\mathbf{T}(1,2) = \frac{1}{z_1 z_2 + z_c(z_1 + z_c + 2z_2)}$$
$$\cdot \begin{bmatrix} \dfrac{z_2 z_c}{2z_1^2 z_2 + z_1^2 z_c + 4z_1 z_2 z_c + 2z_1 z_c^2 - 4z_c^3} & \\ \dfrac{2(z_1 + 2z_c)}{z_1 z_2^2 - 2z_1 z_2 z_c - 2z_1 z_c^2 - 2z_c^3 - 4z_2 z_c^2} & \\ \dfrac{2(2z_2 + z_c)}{z_2 z_c} & \end{bmatrix}. \tag{9.19}$$

9 TRANSMISSION MATRICES FOR CIRCUITS AND DISCRETE MEDIA

The composite transmission matrix can be derived for most circuits having linear time-invariant elements by using the two simple circuits above as building blocks. If the elements are separated electrically by a length of transmission line, the phase shift for that section of line must be included in the recursive equations. The algebra for cascading is time consuming and is best carried out on a digital machine.

2. MULTIMODES IN LINEAR CIRCUITS

The preceding treatment was based on the assumption that only one wave is incident on each side of the two-port network. The more general coupled mode problem has several "waves." Several incident waves are present on each side of the two-port network and must be treated in the analysis.

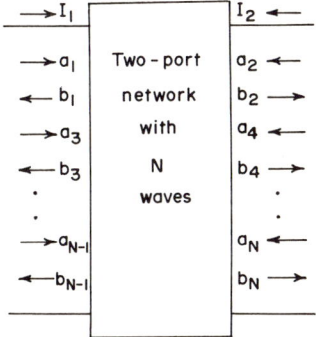

Figure 9.5. Two-port network with N waves.

Assume that there are n waves in the system, half of the waves being incident on one side of the network and the other half striking the opposite side. A network is considered first in which there is no coupling between the waves. It is illustrated in Figure 9.5 as a two-port network with multiple waves. The scattering matrix can be constructed by considering first that the incident waves a_1 and a_2 are of the same type but different from all other incident waves. Similarly, a_3 and a_4 form a pair different from all other pairs. The incident waves a_1 and a_2 can be reflected at the two ports, the reflected waves being b_1 and b_2. It is further assumed that none of the energy from a_1 and a_2 can be coupled into b_i, where $i = 3, 4, \ldots, N$.

The scattering matrix for the multimode two-port network without coupling consists of the individual scattering matrices for each type of wave. The matrix is of order $N \times N$ with the required number of zero entries.

$$\begin{bmatrix} b_1 \\ b_2 \\ b_3 \\ b_4 \\ \cdot \\ \cdot \\ b_{n-1} \\ b_n \end{bmatrix} = \begin{bmatrix} s_{11} & s_{12} & 0 & 0 & 0 & \cdot & \cdot & \cdot \\ s_{21} & s_{22} & 0 & 0 & 0 & \cdot & \cdot & \cdot \\ 0 & 0 & s_{33} & s_{34} & 0 & \cdot & \cdot & \cdot \\ 0 & 0 & s_{43} & s_{44} & 0 & \cdot & \cdot & \cdot \\ 0 & 0 & 0 & 0 & 0 & \cdot & \cdot & \cdot \\ \cdot & \cdot & \cdot & \cdot & \cdot & \cdot & 0 & 0 \\ \cdot & \cdot & \cdot & \cdot & \cdot & 0 & s_{n-1,n-1} & s_{n-1,n} \\ \cdot & \cdot & \cdot & \cdot & \cdot & 0 & s_{n,n-1} & s_{n,n} \end{bmatrix} \begin{bmatrix} a_1 \\ a_2 \\ a_3 \\ a_4 \\ \cdot \\ \cdot \\ a_{n-1} \\ a_n \end{bmatrix} . \quad (9.20)$$

The individual scattering matrices are placed along the diagonal; all other entries are set to zero. This implies, in a mathematical sense, that there is no coupling between a_1 and b_3, etc.

To see how the transmission matrix is formed, consider only the upper left-hand part of the matrix of (9.20), where four waves are considered. The transmission matrix describes the "transmitted" waves rather than the "reflected" waves of the scattering matrix. The first step in finding the **T** matrix for the two-port network is to rearrange (9.20) such that all inputs and outputs are in vector form. If all waves on the left side of the network, are denoted by odd subscripts and those on the right side by even subscripts, the four vectors for the four wave cases are

$$\mathbf{B}_1 = \begin{bmatrix} b_1 \\ b_3 \end{bmatrix}, \quad \mathbf{B}_2 = \begin{bmatrix} b_2 \\ b_4 \end{bmatrix}, \quad \mathbf{A}_1 = \begin{bmatrix} a_1 \\ a_3 \end{bmatrix}, \quad \mathbf{A}_2 = \begin{bmatrix} a_2 \\ a_4 \end{bmatrix}.$$

The scattering matrix for the four waves is then

$$\begin{bmatrix} \mathbf{B}_1 \\ \mathbf{B}_2 \end{bmatrix} = \begin{bmatrix} s_{11} & s_{12} \\ s_{21} & s_{22} \end{bmatrix} \begin{bmatrix} \mathbf{A}_1 \\ \mathbf{A}_2 \end{bmatrix} = \begin{bmatrix} s_{11} & 0 & s_{12} & 0 \\ 0 & s_{33} & 0 & s_{34} \\ s_{21} & 0 & s_{22} & 0 \\ 0 & s_{43} & 0 & s_{44} \end{bmatrix} \begin{bmatrix} a_1 \\ a_3 \\ a_2 \\ a_4 \end{bmatrix}. \quad (9.21)$$

The transformation matrix Ω, given in (2.13) is now applied to the 4×4

scattering matrix of (9.21).

$$[\mathbf{T}] = \begin{bmatrix} s_{12} & 0 & s_{22} & 0 \\ 0 & s_{43} & 0 & s_{44} \\ s_{11} & 0 & s_{12} & 0 \\ 0 & s_{33} & 0 & s_{34} \end{bmatrix}. \tag{9.22}$$

The matrix given in (9.22) is the transmission matrix of the two-port network with two modes. The reader should always keep in mind that the scattering matrix for the vector waves must be of the correct form before applying the transformation.

If there is energy coupling between modes, the zero entries of (9.22) are replaced by the coupling coefficients. This coupling can be between frequency components of a wave, or the coupling may occur between the components of a polarized wave. The derivation of the coupling coefficients is illustrated in the next section.

3. TRANSMISSION MATRIX FOR INTERACTING MODES IN A SERIES-NONLINEAR ELEMENT

When interaction between modes is present, the off-diagonal zeros of (9.21) are replaced by nonzero elements. It is assumed that each mode can produce components of another mode when the medium has the proper characteristics. Consider a two-port network with a nonlinear element which is excited by two electromagnetic waves of frequencies ω_1 and ω_2. If the nonlinear element is a voltage variable capacitor, for example, the two waves mix in the nonlinear element. Harmonics of ω_1 and ω_2 are produced, and these harmonics can produce other waves of the frequencies $m\omega_1 \pm n\omega_2$. The magnitude of the waves of frequencies $m\omega_1 \pm n\omega_2$ decreases when m and n are very large. In most cases the higher-order modes can be ignored since the propagation structure does not permit these modes to propagate. In what follows, only the waves of frequencies ω_1, ω_2, and $\omega_1 + \omega_2$ are considered important and ω_3 is set equal to $\omega_1 + \omega_2$.

The derivation of the scattering matrix cannot be made in a straightforward manner because the transfer coefficients which appear are zero unless two or more modes are present. If we attempt to treat one mode at a time, the interaction is missing and the transfer coefficients are absent. The derivation must be made with all of the modes present, as this is the only way to obtain a valid scattering matrix.

Consider now the circuit shown in Figure 9.6, where it is assumed that ω_3 is present but cannot propagate along the structure. The current for each mode must be continuous across the element; therefore,

$$I_{11} = I_{21}, \tag{9.23a}$$

$$I_{12} = I_{22}, \tag{9.23b}$$

where the first subscript denotes the port and the second subscript the mode. The voltage in each mode must consist of the primary wave and the amount

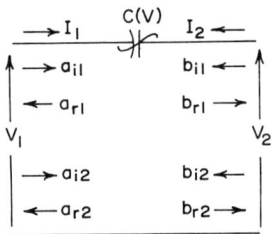

Figure 9.6. Series nonlinear capacitor.

of voltage "transferred" by the nonlinear capacitor. The voltage at port 2 is

$$V_{21} = V_{11} - Z_{11}I_{11} - Z_{12}I_{12}, \tag{9.24a}$$

$$V_{22} = V_{12} - Z_{22}I_{21} - Z_{22}I_{22}, \tag{9.24b}$$

where Z_{jk}, for $j \neq k$, is the "transfer" impedance which generates the additional contribution to the voltage.

Following the general rule of taking the complex conjugate of one mode, the four equations above can be written in a matrix form.

$$\begin{bmatrix} V_{21} \\ V_{22}^* \\ I_{21} \\ I_{22}^* \end{bmatrix} = \begin{bmatrix} 1 & 0 & -Z_{11} & -Z_{12} \\ 0 & 1 & -Z_{21}^* & -Z_{22}^* \\ 0 & 0 & 1 & 0 \\ 0 & 0 & 0 & 1 \end{bmatrix} \begin{bmatrix} V_{11} \\ V_{12}^* \\ I_{11} \\ I_{12}^* \end{bmatrix}. \tag{9.25}$$

The equations are now in the desired form to derive the scattering matrix.

9 TRANSMISSION MATRICES FOR CIRCUITS AND DISCRETE MEDIA

The voltage and current can be written in terms of the modes a_{i1}, a_{i2}, b_{i1}, b_{i2}, etc.,

$$V_{11} = (a_{i1} + a_{r1}), \tag{9.26a}$$

$$I_{11} = (Z_{01})^{-1}(a_{i1} - a_{r1}), \tag{9.26b}$$

where Z_{01} is the characteristic impedance of the line at frequency ω_1. The voltage V_{21}, V_{12}, and V_{22}, as well as the three remaining currents, can be found in a similar way.

All of the voltage and current terms in (9.25) can now be eliminated and the matrix written entirely in terms of the waves and the impedances.

$$\begin{bmatrix} \cdot & (b_{i1} + b_{r1}) \\ \cdot & (b_{i2}^* + b_{r2}^*) \\ (Z_{01})^{-1} & (b_{r1} - b_{i1}) \\ (Z_{02}^*)^{-1} & (b_{r2}^* - b_{i2}^*) \end{bmatrix} = \begin{bmatrix} 1 & 0 & -Z_{11} & -Z_{12} \\ 0 & 1 & -Z_{21}^* & -Z_{22}^* \\ 0 & 0 & 1 & 0 \\ 0 & 0 & 0 & 1 \end{bmatrix} \begin{bmatrix} \cdot & (a_{i1} + a_{r1}) \\ \cdot & (a_{i2}^* + a_{r2}^*) \\ (Z_{01})^{-1} & (a_{i1} - a_{r1}) \\ (Z_{02}^*)^{-1} & (a_{i2}^* - a_{r2}^*) \end{bmatrix}. \tag{9.27}$$

The negative sign on I_{21} and I_{22}^* arises because of the choice of direction assumed for I_2.

The eight mode amplitudes, a_{i1}, a_{r1}, a_{i2}, a_{r2}, b_{i1}, b_{r1}, b_{i2}, and b_{r2}, are analogous to the waves of coupled mode theory. If the matrix is modified so that all of the reflected components can be expressed solely in terms of the impedance matrix and the incident components, the transmission matrix can be found. The matrix of (9.27) can be written as

$$\begin{bmatrix} 1 & 0 & -(1 + Z_{11}Z_{01}^{-1}) & -Z_{12}^* Z_{02}^{*-1} \\ 0 & 1 & -Z_{21}^* Z_{01}^{*-1} & -(1 + Z_{22}Z_{02}^{*-1}) \\ 1 & 0 & 1 & 0 \\ 0 & 1 & 0 & 1 \end{bmatrix} \begin{bmatrix} b_{r1} \\ b_{r2}^* \\ a_{r1} \\ a_{r2}^* \end{bmatrix}$$

$$= \begin{bmatrix} (1 - Z_{11}Z_{01}^{-1}) & -Z_{12}Z_{02}^{*-1} & -1 & 0 \\ -Z_{21}^* Z_{01}^{-1} & (1 - Z_{22}^* Z_{02}^{*-1}) & 0 & -1 \\ 1 & 0 & 1 & 0 \\ 0 & 1 & 0 & 1 \end{bmatrix} \begin{bmatrix} a_{i1} \\ a_{i2}^* \\ b_{i1} \\ b_{i2}^* \end{bmatrix} \tag{9.28}$$

and, finally,

$$\begin{bmatrix} 1 & 0 & -(1+Z_{11}Z_{01}^{-1}) & -Z_{12}^*Z_{02}^{*-1} \\ 0 & 1 & -Z_{21}^*Z_{01}^{-1} & -(1+Z_{22}Z_{02}^{*-1}) \\ 1 & 0 & 1 & 0 \\ 0 & 1 & 0 & 1 \end{bmatrix}^{-1}$$

$$\times \begin{bmatrix} (1-Z_{11}Z_{01}^{-1}) & -Z_{12}Z_{02}^{*-1} & -1 & 0 \\ -Z_{21}^*Z_{01}^{-1} & (1-Z_{22}^*Z_{02}^{*-1}) & 0 & -1 \\ 1 & 0 & 1 & 0 \\ 0 & 1 & 0 & 1 \end{bmatrix}$$

$$= \begin{bmatrix} \tau_{11} & \tau_{12} & R_{11} & R_{12} \\ \tau_{21} & \tau_{22} & R_{21} & R_{22} \\ \mathscr{R}_{11} & \mathscr{R}_{12} & T_{11} & T_{12} \\ \mathscr{R}_{21} & \mathscr{R}_{22} & T_{21} & T_{22} \end{bmatrix}. \quad (9.29)$$

Equation (9.29) is the desired transmission matrix for the lumped nonlinear element. The wave at a frequency ω_3 has been ignored owing to the assumption that it does not propagate. This excitation must be present for the operation of a parametric amplifier, but the method of excitation by ω_3 is arbitrary. If ω_3 is allowed to propagate, then the matrix of (9.29) becomes a 6×6 matrix.

The impedance parameters of (9.29) can be derived in a straightforward manner (see Collin [2]). The impedance matrix for Z is

$$Z = \frac{1}{1-M^2} \begin{bmatrix} 1/j\omega_1 C_0 & -M/j\omega_2 C_0 \\ -M/j\omega_1 C_0 & 1/j\omega_2 C_0 \end{bmatrix}, \quad (9.30)$$

where M is the coupling parameter between ω_1 and ω_2 and C_0 is the static capacitance of the capacitor. The coupling parameter is usually quite small and would never be taken such that $1 - M^2 = 0$.

There may be other methods for obtaining the transmission matrix; the means of deriving the coefficients is left entirely to the reader. The method given here is only for illustrating how he might proceed. The reader should keep in mind the meaning of the transmission matrix when solving the nonlinear element problems. The full set of voltage and currents must be included since the transmission matrix describes the reflection, transmission, and coupling coefficients.

4. TRANSMISSION MATRIX FOR INTERACTING MODES IN A SHUNT-NONLINEAR ELEMENT

The transmission matrix for a shunt element can be derived in a manner similar to that for the series element. If the shunt element is as shown in Figure 9.7, then the voltage across the two-port network must be equal.

$$V_{11} = V_{21}, \tag{9.31a}$$

$$V_{21} = V_{22}. \tag{9.31b}$$

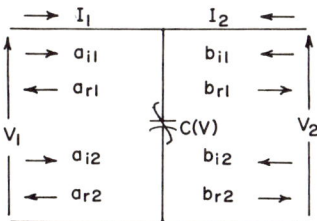

Figure 9.7. Shunt nonlinear capacitor.

The currents in the two-port network are given by

$$I_{21} = -Y_{22}V_{11} - Y_{12}V_{12} + I_{11}, \tag{9.32a}$$

$$I_{22} = -Y_{21}V_{11} - Y_{22}V_{12} + I_{12}, \tag{9.32b}$$

where the subscripts are as given in Section 3. The four equations of (9.31) and (9.32) can now be written in matrix form,

$$\begin{bmatrix} V_{21} \\ V_{22}^* \\ I_{21} \\ I_{22}^* \end{bmatrix} = \begin{bmatrix} 1 & 0 & 0 & 0 \\ 0 & 1 & 0 & 0 \\ -Y_{11} & -Y_{12} & 1 & 0 \\ -Y_{21}^* & -Y_{22}^* & 0 & 1 \end{bmatrix} \begin{bmatrix} V_{11} \\ V_{12}^* \\ I_{11} \\ I_{12}^* \end{bmatrix}, \tag{9.33}$$

where the complex conjugate has been taken for the second and fourth rows.

The transmission matrix for the shunt element can now be found by following the procedure of Section 3. The details are left for the reader to carry out. The transmission matrix is now expressed in terms of admittance parameters since admittance terms appear in (9.33).

5. THE TRANSMISSION MATRIX FOR A LAYERED ISOTROPIC MEDIUM: SINGLE MODE AT NORMAL INCIDENCE

Propagation problems for an isotropic, inhomogeneous medium can be analyzed by using the recursive equations. The propagation of waves in a random medium using the recursive equations was described earlier (see Adams and Denman). All of the necessary equations were given; some of the derivations are repeated here.

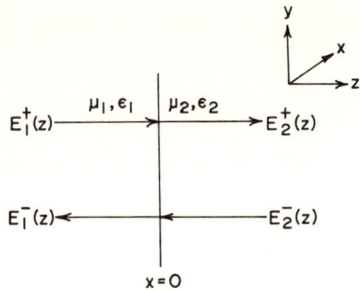

Figure 9.8. Abrupt impedance change.

The single-mode problem is discussed in this section. Several multimode problems are analyzed in later chapters, and the derivation of the multimode transmission matrix is not given here. The examples can serve as a guide for the higher-mode problems which may be encountered.

If an electromagnetic wave of frequency ω strikes an interface dividing two semi-infinite media, the wave is partially transmitted and partially reflected. The characteristic impedance in medium 1 differs from that in medium 2. The wave equation

$$\frac{d^2E(z)}{dz^2} + k^2 E(z) = 0 \qquad (9.34)$$

is assumed to hold in each medium. The two regions of the medium are shown in Figure 9.8. The characteristic impedances for region 1 and region 2 are given by

$$Z_1 = \sqrt{\frac{\mu_1}{\varepsilon_1}}, \qquad Z_2 = \sqrt{\frac{\mu_2}{\varepsilon_2}}$$

and the wave numbers by

$$k_1 = \omega\sqrt{\mu_1 \varepsilon_1}, \qquad k_2 = \omega\sqrt{\mu_2 \varepsilon_2}.$$

9 TRANSMISSION MATRICES FOR CIRCUITS AND DISCRETE MEDIA

To derive the expression for the transmission matrix, use must be made of the continuity of field components at the boundary. If the electric field is aligned along x, then the electric field obeys the equation

$$E_x(z) = E^+ e^{-jkz} + E^- e^{jkz} \tag{9.35a}$$

and the magnetic field is given by

$$H_y(z) = \frac{1}{Z}[E^+ e^{-jkz} - E^- e^{jkz}]. \tag{9.35b}$$

The wave impedance at any point z is given by

$$Z(z) = \frac{E_x(z)}{H_y(z)}. \tag{9.36}$$

The general boundary condition which must hold is that the tangential magnetic and electric fields must be continuous across the boundary.

Using the continuity restriction, then, the equations

$$E_1^+(0-) + E_1^-(0-) = E_2^+(0+) + E_2^-(0+), \tag{9.37a}$$

$$\frac{E_1^+(0-) - E_1^-(0-)}{Z_1} = \frac{E_2^+(0+) - E_2^-(0+)}{Z_2}, \tag{9.37b}$$

must hold at $z = 0$. If there is no incident wave from the right side, $E_2^-(z) = 0$; thus

$$\mathcal{R} = \frac{E_1^-(0-)}{E_1^+(0-)} = \frac{Z_2 - Z_1}{Z_2 + Z_1} \tag{9.38a}$$

and

$$\tau = \frac{E_2^+(0+)}{E_1^+(0-)} = \frac{2Z_2}{Z_2 + Z_1}. \tag{9.38b}$$

If there is no incident wave from the left side, the equations of (9.37) can be manipulated to find R and T.

$$R = \frac{E_2^+(0+)}{E_2^-(0+)} = \frac{Z_1 - Z_2}{Z_1 + Z_2}, \tag{9.38c}$$

$$T = \frac{E_1^-(0-)}{E_2^-(0+)} = \frac{2Z_1}{Z_1 + Z_2}. \tag{9.38d}$$

Equations (9.38) give the four coefficients for a finite impedance step occurring between two semi-infinite media, where the characteristic impedances of the two regions are Z_1 and Z_2.

The four equations can be modified for a medium where the wave number or the refractive index is given. If $\mu_1 \neq \mu_2$, the left-hand coefficients are

$$\mathscr{R} = \frac{\mu_2 k_1 - \mu_1 k_2}{\mu_2 k_1 + \mu_1 k_2}, \tag{9.39a}$$

$$\tau = \frac{2\mu_2 k_1}{\mu_2 k_1 + \mu_1 k_2}. \tag{9.39b}$$

The coefficients for the right side, R and T, can easily be found from (9.38).

If the refractive index is given, where $\eta(x) = \sqrt{\varepsilon(x)}$, the reflection and transmission coefficients take another form. If $\mu_1 = \mu_2$, as is generally true for a material characterized by the refractive index,

$$\mathscr{R} = \frac{\eta_1 - \eta_2}{\eta_1 + \eta_2}, \tag{9.40a}$$

$$\tau = \frac{2\eta_1}{\eta_1 + \eta_2} \tag{9.40b}$$

The remaining two coefficients are not given but can be found directly from (9.40a) and (9.40b) by noting that $\mathscr{R} = -R$ and T is found by interchanging η_1 and η_2 in τ.

The coefficients given are for a single mode with normal incidence on a boundary separating two semi-infinite media. If the medium is broken into layers, the cascading equations given in Chapter 3 are applicable and the composite reflection and transmission coefficients can be found.

6. THE CONTINUOUS MEDIUM AND ELECTROMAGNETIC WAVES AT NORMAL INCIDENCE

Section 5 described the method of finding the transmission matrix for a medium broken into slabs, each slab having a constant impedance. The propagation of electromagnetic waves in this type of medium can be analyzed in an approximate sense only. The slabbing process can give incorrect answers if the slab thickness happens to be a fractional portion of the wavelength. Artificial resonance phenomena can be introduced unless the slab thickness is properly chosen. One way of preventing this resonance is to use a random thickness, but this method is not always satisfactory.

The propagation problem can be analyzed directly by considering the wave equation and the recursive equations for a continuous medium. The wave equation

$$\frac{d^2 E(x)}{dx^2} + k^2(x) E(x) = 0 \tag{9.41}$$

can be analyzed by considering two first-order equations. The two first-order equations are obtained from (9.41). If $dE(x)/dx = v(x)$ and $E(x) = u(x)$, the two equations

$$\frac{du(x)}{dx} = v(x) \tag{9.42a}$$

and

$$-\frac{dv(x)}{dx} = k^2(x) u(x) \tag{9.42b}$$

must be solved. The functions $u(x)$ and $v(x)$ can be calculated to give the field and the first derivative. If the two required boundary conditions are given, the field can be found everywhere.

7. OBLIQUE INCIDENCE CASE

The derivation of the transmission matrix for the oblique incidence case was given in a previously cited reference (Adams and Denman). The equations are not given here because the coefficients are presented in several elementary books on electromagnetic field theory (see Ramo, Whinnery, and Van Duzer [3]).

The oblique incidence case can be analyzed as a multimode problem if the change of polarization of a wave in an inhomogeneous anisotropic medium is to be analyzed. Propagation of an electromagnetic wave in a plasma is discussed in Chapter 12. The coupling between the ordinary and extraordinary wave is described there. The coupled mode method was applied without difficulties.

8. SUMMARY

The derivation of the transmission matrix for the electrical network and for simple electromagnetic problems has been given in this chapter. The methods presented here are not to be taken as the only ways of obtaining the transmission matrix; the work is presented only as a guide to help the reader.

There are numerous ways to describe the transmission matrices; the reader is left to devise the method that will be valid for his particular problem.

REFERENCES

1. J. L. Altman, *Microwave Circuits*, D. Van Nostrand Company, Princeton, N.J., 1964.
2. R. E. Collin, *Foundations for Microwave Engineering*, McGraw-Hill Book Co., New York, 1966.
3. S. Ramo, J. R. Whinnery, and T. Van Duzer, *Fields and Waves in Communications Electronics*, John Wiley and Sons, New York, 1965.

CHAPTER 10

Plane Elastic Waves in an Inhomogeneous Lossless Medium

The propagation of elastic waves through an elastic medium is of interest in the study of geophysics and acoustics. This chapter describes a study of plane elastic waves in a lossless medium, using the technique of invariant imbedding described in earlier chapters. No assertion is made that all problems concerning elastic waves in a material have been solved, nor is it apparent that all problems can be solved using the method. The simplest case of a plane elastic wave in an inhomogeneous lossless medium has been studied and is described herein. There is no obvious reason to believe that many other problems cannot be attacked by this method; the extent to which the method is applicable is not known. The work to be described has been taken from an M.S. thesis completed by one of the author's students, R. E. Dodd, at Vanderbilt University [1].

There are, in general, three basic types of elastic waves that propagate either in the interior or on the free surface of an elastic medium. These waves are: pressure or compressional waves, shear waves, and surface waves. The compressional waves are longitudinal waves that can propagate through any material medium. The longitudinal waves are generally referred to as the P waves. The shear waves, of which there are two types, are transverse waves which are restricted to solid media and nonideal fluids. Since these waves are shear waves, an ideal fluid cannot support the propagation of them. If the shear waves have a vertical polarization, the waves are normally designated SV waves. Shear waves with horizontal polarization are designated SH waves. The P and SV waves are considered in this chapter, although the SH wave could be included by enlarging the order of the matrix system.

The surface waves occur in bounded media; the work here is for unbounded media. The surface waves are neglected, in part, because there has been no attempt to consider how to include them in the invariant imbedding formulation. This area of research is one that needs to receive some attention before a complete understanding of elastic wave propagation is available.

The medium to be considered is inhomogeneous in one dimension, along the direction of propagation. The medium is a layered medium consisting of semi-infinite layers in the x-y plane, each layer being homogeneous and isotropic but differing from adjacent layers. The discrete model of invariant imbedding can be applied and the reflection, transmission, and coupling of P and SV waves studied. The extension to the continuous case is not difficult, although few materials exist that are continuously inhomogeneous.

To apply the invariant imbedding technique the interaction of the P and SV waves at the simple interfaces are considered. The shear and pressure waves propagate independently of each other in the homogeneous slabs, each wave undergoing a phase shift in the slab. Since the material is assumed to be lossless, there is no damping of the wave in the slab. Absorption could easily be considered in the formulation.

I. PROPAGATION IN AN UNBOUNDED MEDIUM

Before the numerical method of invariant imbedding can be applied, the reflection, transmission, and coupling coefficients at a simple interface must be found. Before giving the equations for finding these coefficients, the fundamentals of elasticity are discussed briefly.

When an external force is applied to an elastic body, the body may undergo a translation, a rotation, or a change of shape. Each point in the strained body experiences a displacement from its position in the unstrained body. Let the point $P(x, y, z)$ undergo a displacement $\bar{s} = u\bar{a}_x + v\bar{a}_y + w\bar{a}_z$, where u, v, and w are the components of displacement along the x, y, and z axes, respectively. After deformation the coordinates of P become $(x + u, y + v, z + w)$, where it is assumed that u, v and w are small. If we consider a point $Q(x + \Delta x, y + \Delta y, z + \Delta z)$ in the neighborhood of P, we see that it undergoes a displacement $(u + \Delta u, v + \Delta v, w + \Delta w)$. The relative change in displacement $(\Delta u, \Delta v, \Delta w)$ of Q can then be found from a Taylor series expansion:

$$\Delta u = \frac{\partial u}{\partial x}\Delta x + \frac{\partial u}{\partial y}\Delta y + \frac{\partial u}{\partial z}\Delta z, \tag{10.1a}$$

$$\Delta v = \frac{\partial v}{\partial x}\Delta x + \frac{\partial v}{\partial y}\Delta y + \frac{\partial v}{\partial z}\Delta z, \tag{10.1b}$$

$$\Delta w = \frac{\partial w}{\partial x}\Delta x + \frac{\partial w}{\partial y}\Delta y + \frac{\partial w}{\partial z}\Delta z, \tag{10.1c}$$

where higher-order terms are neglected.

10 PLANE ELASTIC WAVES IN AN INHOMOGENEOUS LOSSLESS MEDIUM

Now let

$$\varepsilon_{xx} = \frac{\partial u}{\partial x}, \qquad \varepsilon_{yy} = \frac{\partial v}{\partial y}, \qquad \varepsilon_{zz} = \frac{\partial w}{\partial z} \qquad (10.2a)$$

and

$$\varepsilon_{xy} = \frac{\partial v}{\partial x} + \frac{\partial u}{\partial y}, \qquad \varepsilon_{yz} = \frac{\partial w}{\partial y} + \frac{\partial v}{\partial z}, \qquad \varepsilon_{zx} = \frac{\partial u}{\partial z} + \frac{\partial w}{\partial x}, \qquad (10.2b)$$

where ε_{xx}, ε_{yy} and ε_{zz} correspond to fractional changes in length along the x, y, and z axes, respectively. The quantities ε_{xy}, ε_{yz}, and ε_{zx} represent the shear strains in the plane denoted by their subscripts. Let Ω_x, Ω_y, and Ω_z denote components of rotation, where

$$2\Omega_x = \frac{\partial w}{\partial y} - \frac{\partial v}{\partial z}, \qquad 2\Omega_y = \frac{\partial u}{\partial z} - \frac{\partial w}{\partial x}, \qquad 2\Omega_z = \frac{\partial v}{\partial x} - \frac{\partial u}{\partial y}. \qquad (10.2c)$$

Equation (10.1) can then be written

$$\Delta u = [\varepsilon_{xx} \Delta x + \tfrac{1}{2}\varepsilon_{xy} \Delta y + \tfrac{1}{2}\varepsilon_{zx} \Delta z] + [\Omega_y \Delta z - \Omega_z \Delta y], \qquad (10.3a)$$

$$\Delta v = [\tfrac{1}{2}\varepsilon_{xy} \Delta x + \varepsilon_{yy} \Delta y + \tfrac{1}{2}\varepsilon_{zy} \Delta z] + [\Omega_z \Delta x - \Omega_x \Delta x], \qquad (10.3b)$$

$$\Delta w = [\tfrac{1}{2}\varepsilon_{xz} \Delta x + \tfrac{1}{2}\varepsilon_{yz} \Delta y + \varepsilon_{zz} \Delta x] + [\Omega_x \Delta y - \Omega_y \Delta x], \qquad (10.3c)$$

where symmetry has been assumed.

The strains in an elastic body occur as a result of the forces distributed throughout the body. The ratio of the force acting on an element of area S to the area is defined as the stress. In general, the stress on a surface element does not act normal to the surface; therefore it is convenient to resolve the stress into normal and tangential components at the surface. If the stress components are designated by P_{mn}, where m refers to the coordinate axis along which the component of stress is directed and n the plane on which it acts, then there are nine stress components. If the body on which the stress acts is to be in equilibrium, there must be symmetry for three components, $P_{xy} = P_{yx}$, $P_{xz} = P_{zx}$, and $P_{yz} = P_{zy}$. The stress can therefore be defined by six components: P_{xx}, P_{yy}, and P_{zz} are tensional or compressive stresses depending on the sign of the components, and P_{xy}, P_{yz}, and P_{zx} are shearing stresses which act on the body.

The six components of stress at any point of an elastic body can be expressed as a linear function of six components of strain, provided that the strain components are small. The stress-strain relations can be written in terms of Lame's constant λ and μ for an isotropic solid. The stress-strain relations are

$$P_{xx} = (\lambda + 2\mu)\varepsilon_{xx} + \lambda\varepsilon_{yy} + \lambda\varepsilon_{zz}, \tag{10.4a}$$

$$P_{yy} = \lambda\varepsilon_{xx} + (\lambda + 2\mu)\varepsilon_{yy} + \lambda\varepsilon_{zz}, \tag{10.4b}$$

$$P_{zz} = \lambda\varepsilon_{xx} + \lambda\varepsilon_{yy} + (\lambda + 2\mu)\varepsilon_{zz}, \tag{10.4c}$$

$$P_{xy} = \mu\varepsilon_{xy}, \tag{10.4d}$$

$$P_{yz} = \mu\varepsilon_{yz}, \tag{10.4e}$$

$$P_{zx} = \mu\varepsilon_{zx}, \tag{10.4f}$$

where

$$\lambda = \frac{\sigma E}{(1+\sigma)(1-2\sigma)}, \quad \mu = \frac{E}{2(1+\sigma)}, \tag{10.5a}$$

$$E = \frac{\mu(3\lambda + 2\mu)}{\lambda + \mu}, \quad \sigma = \frac{\lambda}{2(\lambda + \mu)}, \quad \kappa = \lambda + \tfrac{2}{3}\mu, \tag{10.5b}$$

and E is Young's modulus, σ is Poisson's ratio, and κ is the bulk modulus of the material.

The equations of motion can now be obtained from the expressions above. Consider an infinitesimal element of volume in the form of a parallelepiped with sides parallel to the coordinate planes; see Figure 10.1. Three stresses act on each face, and there is a total of eighteen stresses for the entire volume. Six separate forces therefore act on the parallelepiped, and these forces are parallel to the axes. The forces in the x direction are

$$\left(P_{xx} + \frac{\partial P_{xx}}{\partial x}\Delta x\right)\Delta y\, \Delta z - P_{xx}\Delta y\, \Delta z$$

$$+ \left(P_{xy} + \frac{\partial P_{xy}}{\partial y}\Delta y\right)\Delta x\, \Delta z - P_{xy}\Delta x\, \Delta z$$

$$+ \left(P_{xz} + \frac{\partial P_{xz}}{\partial z}\Delta z\right)\Delta x\, \Delta y - P_{xz}\Delta x\, \Delta y = \Delta F_x, \tag{10.6}$$

which reduces to

$$\left(\frac{\partial P_{xx}}{\partial x} + \frac{\partial P_{xy}}{\partial y} + \frac{\partial P_{xz}}{\partial z}\right)\Delta x\, \Delta y\, \Delta z = \Delta F_x. \tag{10.7}$$

10 PLANE ELASTIC WAVES IN AN INHOMOGENEOUS LOSSLESS MEDIUM

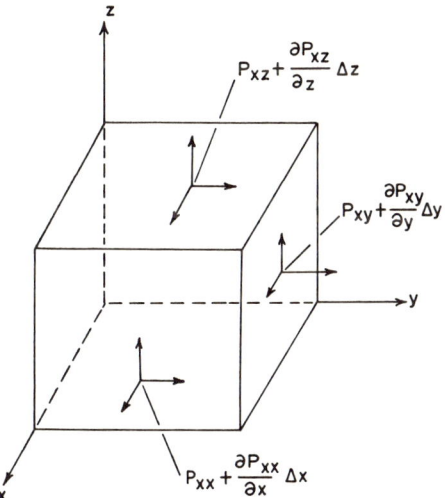

Figure 10.1. Forces acting on a parallelepiped.

Applying Newton's second law to the volume, we find

$$\rho \, \Delta x \, \Delta y \, \Delta z \frac{\partial^2 u}{\partial t^2} = \left(\frac{\partial P_{xx}}{\partial x} + \frac{\partial P_{xy}}{\partial y} + \frac{\partial P_{xz}}{\partial z} \right) \Delta x \, \Delta y \, \Delta z + \rho X \, \Delta x \, \Delta y \, \Delta z. \tag{10.8}$$

The force acting along the x axis is

$$\rho \frac{\partial^2 u}{\partial t^2} = \frac{\partial P_{xx}}{\partial x} + \frac{\partial P_{xy}}{\partial y} + \frac{\partial P_{xz}}{\partial z} + \rho X, \tag{10.9a}$$

where X is the body force per unit mass. The forces along y and z are found in a similar way, and the result is

$$\rho \frac{\partial^2 v}{\partial t^2} = \frac{\partial P_{yx}}{\partial x} + \frac{\partial P_{yy}}{\partial y} + \frac{\partial P_{yz}}{\partial z} + \rho Y, \tag{10.9b}$$

$$\rho \frac{\partial^2 w}{\partial t^2} = \frac{\partial P_{zx}}{\partial x} + \frac{\partial P_{zy}}{\partial y} + \frac{\partial P_{zz}}{\partial z} + \rho Z, \tag{10.9c}$$

where ρ is the density of the material and X, Y, and Z are the body forces, which are neglected.

The equations of (10.9) can now be rewritten in terms of the particle displacements (u, v, w) by using (10.2).

$$\rho \frac{\partial^2 u}{\partial t^2} = (\lambda + 2\mu) \frac{\partial^2 u}{\partial x^2} + \mu \left(\frac{\partial^2 u}{\partial y^2} + \frac{\partial^2 u}{\partial z^2} \right) + (\lambda + \mu) \left(\frac{\partial^2 v}{\partial x \, \partial y} + \frac{\partial^2 w}{\partial x \, \partial z} \right),$$
(10.10a)

$$\rho \frac{\partial^2 v}{\partial t^2} = (\lambda + 2\mu) \frac{\partial^2 v}{\partial y^2} + \mu \left(\frac{\partial^2 v}{\partial x^2} + \frac{\partial^2 v}{\partial z^2} \right) + (\lambda + \mu) \left(\frac{\partial^2 u}{\partial x \, \partial y} + \frac{\partial^2 w}{\partial y \, \partial z} \right),$$
(10.10b)

$$\rho \frac{\partial^2 w}{\partial t^2} = (\lambda + 2\mu) \frac{\partial^2 w}{\partial z^2} + \mu \left(\frac{\partial^2 w}{\partial x^2} + \frac{\partial^2 w}{\partial y^2} \right) + (\lambda + \mu) \left(\frac{\partial^2 u}{\partial x \, \partial z} + \frac{\partial^2 v}{\partial y \, \partial z} \right).$$
(10.10c)

Using the Laplacian operator and the relation

$$\Theta = \varepsilon_{xx} + \varepsilon_{yy} + \varepsilon_{zz} = \partial u/\partial x + \partial v/\partial y + \partial w/\partial z,$$

we finally obtain

$$\rho \frac{\partial^2 u}{\partial t^2} = (\lambda + \mu) \frac{\partial \Theta}{\partial x} + \mu \nabla^2 u, \qquad (10.11a)$$

$$\rho \frac{\partial^2 v}{\partial t^2} = (\lambda + \mu) \frac{\partial \Theta}{\partial y} + \mu \nabla^2 v, \qquad (10.11b)$$

$$\rho \frac{\partial^2 w}{\partial t^2} = (\lambda + \mu) \frac{\partial \Theta}{\partial z} + \mu \nabla^2 w, \qquad (10.11c)$$

where Θ is referred to as the dilatation of the volume.

The three equations of (10.11) define two types of elastic wave propagation in a material. The dilatation wave obeys the wave equation

$$\nabla^2 \Theta = \frac{\rho}{\lambda + 2\mu} \frac{\partial^2 \Theta}{\partial t^2} \qquad (10.12)$$

obtained by differentiating (10.11a), (10.11b), (10.11c) with respect to x, y, and z, respectively, and then adding the three resulting equations. The dilatation wave is the compression or P wave in the medium.

The rotational or shear wave equations can be found by differentiating (10.11b) with respect to z and (10.11c) with respect to y and subtracting the latter from the former; similar manipulations are used for the other

10 PLANE ELASTIC WAVES IN AN INHOMOGENEOUS LOSSLESS MEDIUM

combinations of the equations of (10.11).

$$\nabla^2 \Omega_x = \frac{\rho}{\mu} \frac{\partial^2 \Omega_x}{\partial t^2}, \tag{10.13a}$$

$$\nabla^2 \Omega_y = \frac{\rho}{\mu} \frac{\partial^2 \Omega_y}{\partial t^2}, \tag{10.13b}$$

$$\nabla^2 \Omega_z = \frac{\rho}{\mu} \frac{\partial^2 \Omega_z}{\partial t^2}. \tag{10.13c}$$

To find the reflection, transmission, and coupling coefficients at an interface, the vector and scalar potential functions are needed. Let the scalar potential ϕ be defined such that

$$u = \frac{\partial \phi}{\partial x}, \quad v = \frac{\partial \phi}{\partial y}, \quad w = \frac{\partial \phi}{\partial z}; \tag{10.14}$$

then the wave equation for ϕ is given by

$$\nabla^2 \phi = \frac{1}{C_L^2} \frac{\partial^2 \phi}{\partial t^2}, \tag{10.15a}$$

where

$$C_L^2 = \frac{\lambda + 2\mu}{\rho}. \tag{10.15b}$$

This wave equation is the same as (10.12); thus we can conclude that the function ϕ is a valid scalar function for the compressional wave.

The vector function ψ is obtained by defining

$$u = \frac{\partial \psi_z}{\partial y} - \frac{\partial \psi_y}{\partial z}, \quad v = \frac{\partial \psi_x}{\partial z} - \frac{\partial \psi_z}{\partial x}, \quad w = \frac{\partial \psi_y}{\partial x} - \frac{\partial \psi_x}{\partial y}, \tag{10.16}$$

where ψ_m satisfies the wave equation

$$\nabla^2 \psi_m = \frac{1}{C_t^2} \frac{\partial^2 \psi_m}{\partial t^2}, \tag{10.17a}$$

with $m = x, y,$ or z. This wave equation is the same as (10.13) and can be used to describe the shear wave. The velocity C_t is defined through the equation

$$C_t^2 = \frac{\mu}{\rho}. \tag{10.17b}$$

The boundary conditions imposed on elastic waves at any interface involve the continuity of the normal and tangential displacements and stresses. Using the potential functions, we find from the equations given above that the displacement components in a solid are given by

$$u = \frac{\partial \phi}{\partial x} + \frac{\partial \psi_z}{\partial y} - \frac{\partial \psi_y}{\partial z}, \tag{10.18a}$$

$$v = \frac{\partial \phi}{\partial y} + \frac{\partial \psi_x}{\partial z} - \frac{\partial \psi_z}{\partial x}, \tag{10.18b}$$

$$w = \frac{\partial \phi}{\partial z} + \frac{\partial \psi_y}{\partial x} - \frac{\partial \psi_z}{\partial y}. \tag{10.18c}$$

If a P and an SV wave are the only waves considered, then the three equations reduce to only two equations:

$$u = \frac{\partial \phi}{\partial x} - \frac{\partial \psi_y}{\partial z}, \tag{10.19a}$$

$$w = \frac{\partial \phi}{\partial z} + \frac{\partial \psi_y}{\partial x}. \tag{10.19b}$$

The normal and tangential stress equations at any plane which is defined by z equal to a constant, when the SH wave is absent, are given by

$$P_{zz} = \lambda \left(\frac{\partial^2 \phi}{\partial x^2} + \frac{\partial^2 \phi}{\partial z^2} \right) + 2\mu \left(\frac{\partial^2 \phi}{\partial z^2} + \frac{\partial^2 \psi}{\partial x\, \partial z} \right), \tag{10.20a}$$

$$P_{zx} = \mu \left(2 \frac{\partial^2 \phi}{\partial z\, \partial x} - \frac{\partial^2 \psi}{\partial z^2} + \frac{\partial^2 \psi}{\partial x^2} \right). \tag{10.20b}$$

It is now a simple matter to obtain the reflection, transmission, and coupling coefficients at a simple interface.

2. INTERFACE COEFFICIENTS

Consider two semi-infinite solid media separated by an interface which is parallel to the x-y plane. Let a pressure and a shear wave be incident on the interface and let the angles of incidence be given by θ_p and θ_s, respectively; see Figure 10.2. The material constants are μ, λ, ρ, k, and K, where k and K are the wave numbers for the P and SV waves, respectively. All constants and functions are described by the primes in the mth medium, unprimed for the $(m-1)$th medium.

10 PLANE ELASTIC WAVES IN AN INHOMOGENEOUS LOSSLESS MEDIUM

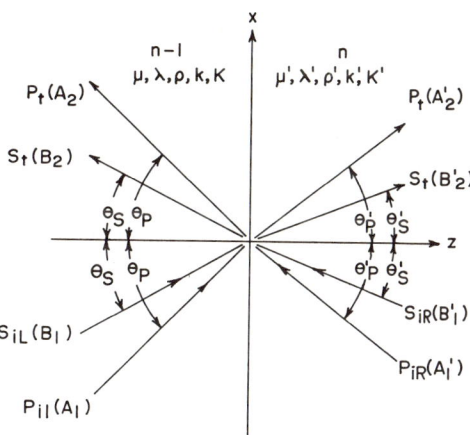

Figure 10.2. Interaction of P and SV waves at a solid-solid interface.

The potential functions for the left and right sides of Figure 10.2 are given by

$$\phi_L = \{A_1 \exp[-jk(x \sin \theta_p + z \cos \theta_p)] \\ + A_2 \exp[jk(-x \sin \theta_p + z \cos \theta_p)]\}\exp(j\omega t), \quad (10.21a)$$

$$\phi_R = \{A_1' \exp[jk'(-x \sin \theta_p' + z \cos \theta_p')] \\ + A_2' \exp[-jk'(x \sin \theta_p' + z \cos \theta_p')]\}\exp(j\omega t), \quad (10.21b)$$

$$\psi_L = \{B_1 \exp[-jK(x \sin \theta_s + z \cos \theta_s)] \\ + B_2 \exp[jK(-x \sin \theta_s + z \cos \theta_s)]\}\exp(j\omega t), \quad (10.21c)$$

$$\psi_R = \{B_1' \exp[jK'(-x \sin \theta_s' + z \cos \theta_s')] \\ + B_2' \exp[-jK'(x \sin \theta_s' + z \cos \theta_s')]\}\exp(j\omega t), \quad (10.21d)$$

where

A_1, A_1' = amplitudes of incident P waves,
B_1, B_1' = amplitudes of incident S waves,
A_2, A_2' = amplitudes of transmitted P waves,
B_2, B_2' = amplitudes of transmitted S waves,
θ_p, θ_p' = angle of incidence of the P wave,
θ_s, θ_s' = angle of incidence of the S wave,
ω = angular frequency,
t = time,
$j = \sqrt{-1}$.

The incident wave is assumed to be polarized in the vertical direction so that there is no y component of the waves (SH component).

The sixteen reflection, transmission, and coupling coefficients can now be found for the solid-solid interface. Since this is a two-mode problem there are four coefficients for each incident wave, and therefore a total of sixteen coefficients for the four incident waves. To find the sixteen coefficients the boundary conditions are applied for each incident wave when the other three incident waves are absent. For a solid-solid interface the boundary conditions require continuity of the normal and tangential displacements and stresses at the interface; this means that: $u = u'$, $w = w'$, $P_{zz} = P'_{zz}$, and $P_{zx} = P'_{zx}$ at $z = 0$. The potentials ϕ and ψ must be continuous across the interface; this means that the exponential terms evaluated at the interfaces must be equal and implies that

$$k \sin \theta_p = k' \sin \theta_p' = K \sin \theta_s = K' \sin \theta_s', \qquad (10.22)$$

which is Snell's law.

Applying the boundary conditions and using (10.22) to simplify, we find that

$$A_1 + A_2 + b(B_2 - B_1) = A_1' + A_2' + b'(B_1' - B_2') \quad \text{for } u = u', \quad (10.23a)$$

$$a(A_1 - A_2) + B_1 + B_2 = a'(A_2' - A_1') + (B_1' + B_2') \quad \text{for } w = w', \qquad (10.23b)$$

$$\lambda\left\{\left(1 + a^2 + \frac{2\mu a^2}{\lambda}\right)(A_1 + A_2) + \frac{2\mu b}{\lambda}(B_1 - B_2)\right\}$$
$$= \lambda'\left\{\left(1 + a'^2 + \frac{2\mu' a'^2}{\lambda'}\right)(A_1' + A_2') + \frac{2\mu' b'}{\lambda'}(B_2' - B_1')\right\} \quad \text{for } P_{zz} = P'_{zz}, \qquad (10.23c)$$

$$\mu\{2a(A_1 - A_2) + (1 - b^2)(B_1 + B_2)\}$$
$$= \mu'\{-2a'(A_1' - A_2')\} \quad \text{for } P_{zx} = P'_{zx}, \quad (10.23d)$$

where

$$a = \cot \theta_p, \qquad a' = \cot \theta_p', \qquad b = \cot \theta_s, \qquad b' = \cot \theta_s'.$$

The equations above are general, since all four waves are considered to be incident on the interface. Consider the reduced equations when the P wave is incident from the left side by letting $A_1' = B_1 = B_1' = 0$. The equations for an incident P wave are

$$A_1 + A_2 + bB_2 = A_2' - b'B_2', \quad (10.24a)$$

$$a(A_1 + A_2) + B_2 = a'A_2' + B_2', \quad (10.24b)$$

$$\lambda\left\{\left(1 + a^2 + \frac{2\mu a^2}{\lambda}\right)(A_1 + A_2) - \frac{2\mu b}{\lambda} B_2\right\}$$
$$= \lambda'\left\{\left(1 + a'^2 + \frac{2\mu' a'^2}{\lambda'}\right)A_2' + \frac{2\mu' b'}{\lambda'} B_2'\right\}, \quad (10.24c)$$

$$\mu\{2a(A_1 - A_2) + (1 - b^2)B_2\} = \mu'\{2a'A_2' + (1 - b'^2)B_2'\}. \quad (10.24d)$$

The four reflection, transmission, and coupling coefficients can now be found for this case. Dividing each equation by A_1 and defining the coefficients

$$t_{11}^p = \frac{A_2'}{A_1}, \quad k_{11}^{sp} = \frac{B_2'}{A_1}, \quad r_{21}^p = \frac{A_2}{A_1}, \quad k_{21}^{sp} = \frac{B_2}{A_1},$$

we obtain the desired final forms for the equations:

$$1 = t_{11}^p - b'k_{11}^{sp} - r_{21}^p - bk_{21}^{sp}, \quad (10.25a)$$

$$1 = \frac{a'}{a} t_{11}^p + \frac{1}{a} k_{11}^{sp} + r_{21}^p - \frac{1}{a} k_{21}^{sp}, \quad (10.25b)$$

$$1 = \frac{\lambda'(1 + a'^2 + 2\mu' a'/\lambda')}{\lambda(1 + a^2 + 2\mu a^2/\lambda)} t_{11}^p + \frac{2\mu' b'}{\lambda(1 + a^2 + 2\mu a^2/\lambda)} k_{11}^{sp} - r_{21}^p$$
$$+ \frac{2\mu b}{\lambda(1 + a^2 + 2\mu a^2/\lambda)} k_{21}^{sp}, \quad (10.25c)$$

$$1 = \frac{\mu' a'}{\mu a} t_{11}^p + \frac{\mu'(1 - b'^2)}{2\mu a} k_{11}^{sp} + r_{21}^p - \frac{(1 - b^2)}{2a} k_{21}^{sp}. \quad (10.25d)$$

Four equations are now available for finding the four coefficients: t_{11}^p is the transmission coefficient for the P wave, k_{11}^{sp} is the coupling coefficient between an incident pressure and a transmitted shear wave, r_{21}^p is the reflection coefficient for the pressure wave, and k_{21}^{sp} is the coupling coefficient for an incident pressure wave reflected as a shear wave. These four coefficients can be calculated for each interface and then be used in the recursive equations as required.

The remaining twelve coefficients for the solid-solid interface can be obtained from (10.24) by letting other waves be incident on the interface. The details of the equations are omitted since their solution is a simple matter of solving algebraic equations. The resulting sixteen coefficients are programmed for the digital machine and solved as part of the overall problem.

If an ideal liquid-solid or solid-liquid interface occurs, the potential functions must be modified. Since an ideal liquid cannot support the propagation of shear waves, the potential functions must have this component set equal to zero. The details of this derivation are left to the reader. No new equations are required, the only change being in the potential functions.

A set of general equations for the three types of interfaces discussed above are contained in the thesis by Dodd [1] (copies of the thesis can be obtained).

3. CONSERVATION OF ENERGY

It is always useful in numerical calculations to have a checking procedure in the program to determine if the computations are well-behaved. The check also provides a means of verifying the equations which are used. For the lossless material the condition of conservation of energy provides a means of checking both the program and the equations.

Ewing, Jardetsky, and Press [2] have derived equations which relate the transmission, reflection, and coupling of energy that occurs at an interface as a result of an incident P or an S wave. These equations were obtained by equating the energy in the incident wave to the sum of the energy of the reflected and transmitted waves of both types and normalizing the incident energy to unity.

For a P wave incident from the left,

$$1 = (r_{21}^p)^2 + \frac{b}{a}(k_{21}^{sp})^2 + \frac{\rho'a'}{\rho a}(t_{11}^p)^2 + \frac{\rho'b'}{\rho a}(k_{11}^{sp})^2, \qquad (10.26a)$$

and, for the S wave incident from the left,

$$1 = \frac{a}{b}(k_{21}^{ps})^2 + (r_{21}^s)^2 + \frac{\rho'a'}{\rho b}(k_{11}^{ps})^2 + \frac{\rho'b'}{\rho b}(t_{11}^s)^2. \qquad (10.26b)$$

A similar set of equations can be written for waves incident from the right side of the interface. The equation for the energy of the P wave is

$$1 = (r_{12}^p)^2 + \frac{b'}{a'}(k_{12}^{sp})^2 + \frac{\rho a}{\rho'a'}(t_{22}^p)^2 + \frac{\rho b}{\rho'a'}(k_{22}^{sp})^2 \qquad (10.26c)$$

and, for the S wave,

$$1 = \frac{a'}{b'}(k_{12}^{ps})^2 + (r_{12}^s)^2 + \frac{\rho a}{\rho'b'}(k_{22}^{ps})^2 + \frac{\rho b}{\rho'b'}(t_{22}^s)^2. \qquad (10.26d)$$

10 PLANE ELASTIC WAVES IN AN INHOMOGENEOUS LOSSLESS MEDIUM

Although the equations above were derived under the assumption of conservation of energy across an interface, the equations are valid for a lossless medium when the reflection, transmission, and coupling coefficients are the composite coefficients.

4. NUMERICAL RESULTS

The propagation of P and SV waves in layered media was studied by investigating several models. The numerical results of the study are now given.

Consider an inhomogeneous medium formed by placing a slab of sandstone of thickness Δs in an infinite medium of water. The water is on both

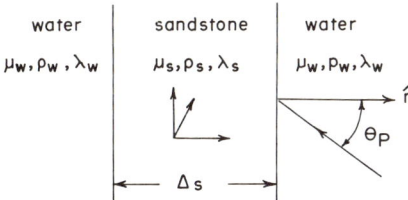

Figure 10.3. Two-interface elastic medium.

sides of the sandstone as shown in Figure 10.3. The drawing shows $\hat{n}$ as the unit normal vector to the surface of the sandstone and θ_p as the angle of incidence of the pressure wave. No shear wave can exist in the water, and only two interfaces of the solid-liquid type need to be considered.

The elastic constants and densities for the different layers shown in Figure 10.3 are available in the literature. The necessary constants for water, sandstone and limestone media are given in Table 10.1.

The medium is symmetric; therefore only two of the four recursive equations are needed. If the medium is symmetric, the equations for $R(1, 2, \ldots, j)$ and $T(1, 2, \ldots, j)$ can be solved without considering the remaining two recursive equations. The four interface reflection and transmission coefficients are required; they are found from the equations of

Table 10.1. Elastic Constants for Water, Sandstone, and Limestone

Material	μ (dynes/cm)	ρ (gm/cm)	λ (dynes/cm)
Water	0.0	1.0	2.25×10^{11}
Sandstone	0.6×10^{11}	2.3	0.8×10^{11}
Limestone	2.5×10^{11}	2.7	3.0×10^{11}

164   COUPLED MODES IN PLASMAS, ELASTIC MEDIA, PARAMETRIC AMPLIFIERS

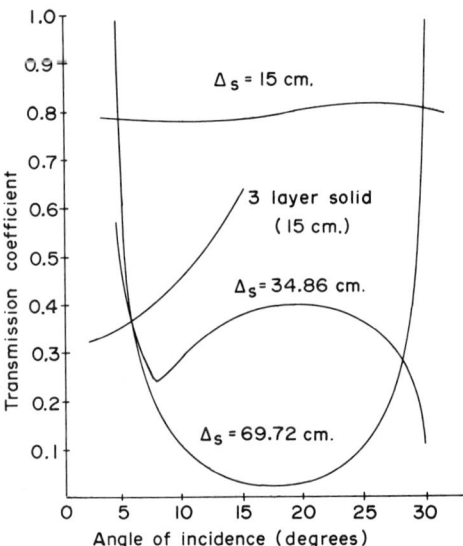

Figure 10.4. Transmission coefficients for varying angles of incidence.

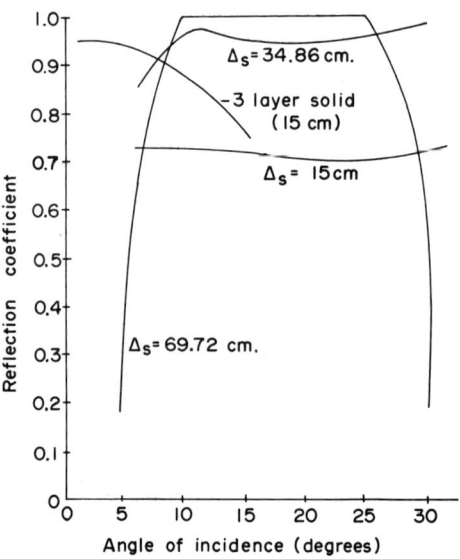

Figure 10.5. Reflection coefficients for varying angles of incidence.

Section 2. It can be shown that $\mathscr{R}(1, 2, \ldots, j)$ is dependent on $T(1, 2, \ldots, j)$ and $\tau(1, 2, \ldots, j)$; thus the incident wave should be always assumed to be incident on the right side for a symmetrical medium.

Before reporting on the data obtained from the investigation of the layered model of Figure 10.3, a few comments on the angular dependency in the medium are in order. If θ_p is the angle of incidence for the wave at the right side, the angle of incidence for the left interface is different. Since the recursive equations can be written for the medium increasing in either direction, the medium thickness could increase from right to left. If the medium is "built up" by moving from right to left, the recursive equations are modified in such a way that $R(1, 2, \ldots, j)$ and $T(1, 2, \ldots, j)$ are no longer independent of $\tau(1, 2, \ldots, j)$ and $\mathscr{R}(1, 2, \ldots, j)$. Rather than proceeding in this manner, the calculations were made by increasing the medium thickness from left to right. Since the medium is symmetric, the angle of incidence for the P and SV waves on the left interface can easily be found from Snell's law. If the medium is not symmetrical and is inhomogeneous, the angles can be found by proceeding from right to left, calculating θ_p and θ_s at each interface until the first interface is reached. The angle at the first interface can then be used to start the calculations.

The numerical method was applied to the lossless model of Figure 10.3, where the sandstone thickness was varied from 0.5 Λ_p to 1.0 Λ_p, Λ_p being the wavelength of the P wave in the sandstone. The frequency of the P and SV waves was taken as 2388 cycles per second. The curves presented in Figure 10.4 and 10.5 show the behavior of the transmission and reflection coefficients for various angles of incidence and sandstone thickness. Total reflection occurs when the sandstone thickness is 0.585 Λ_p and the angle of incidence is approximately 17.5°. The wavelength of the P wave in the sandstone was 123.5 cm for the calculations above.

Figures 10.6 and 10.7 illustrate how the reflection and transmission coefficients vary for the one-layer medium as the thickness of the slab is

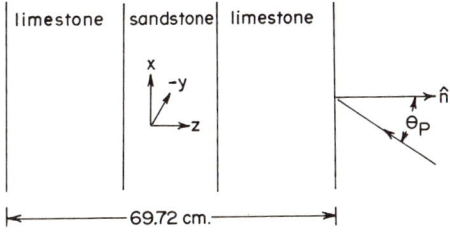

Figure 10.6. Three-layer solid.

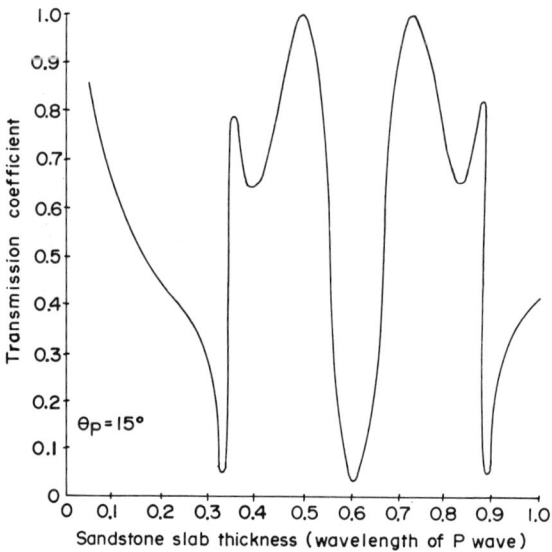

Figure 10.7. Transmission coefficients for varying sandstone slab thickness.

varied. The transmission and reflection coefficients show symmetry about a thickness of approximately $0.6\ \Lambda_p$. The curves are not exact and, owing to the difficulty in drawing the curves, some asymmetry is shown in them. The P wave is incident on the medium at an angle of $15°$.

The model analyzed above is a simple liquid-solid-liquid medium. This particular model has only two interfaces, and the SV wave does not impinge on a solid-solid interface. To study the behavior of the propagation of the P wave when the SV and P waves undergo internal reflections, the model shown in Figure 10.8 was studied. A limestone layer was sandwiched between two layers of sandstone. The limestone and sandstone layers were varied in thickness in such a way as to keep the overall structure of constant thickness. The overall thickness of the three-layered medium was fixed at 69.72 cm.

The data obtained from the calculations are shown in Figures 10.4, 10.5, 10.9, and 10.10. The curves labeled "3-layer solid (15 cm)" in Figures 10.4 and 10.5 are the reflection and transmission coefficients for various incidence angles. The limestone thickness, for each layer, was 15 cm; the sandstone thickness was 39.72 cm.

The transmission and reflection coefficients for varying thickness of the three-layered medium are shown in Figures 10.9 and 10.10. The incident

10 PLANE ELASTIC WAVES IN AN INHOMOGENEOUS LOSSLESS MEDIUM

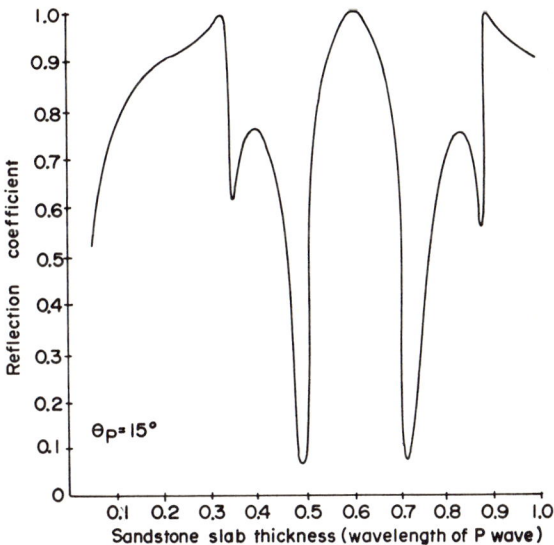

Figure 10.8. Reflection coefficients for varying sandstone slab thickness.

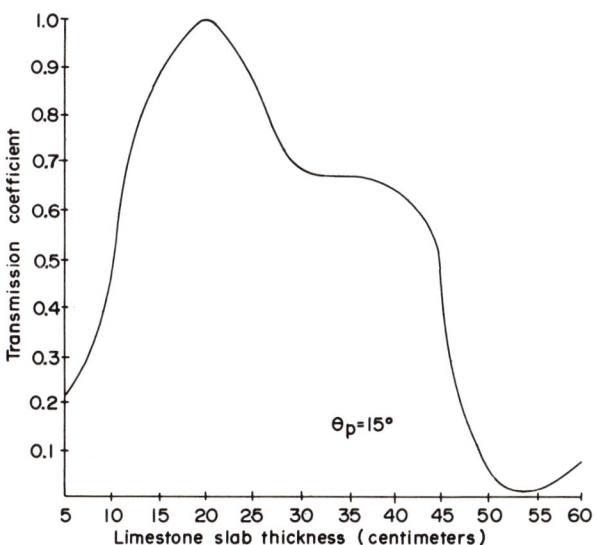

Figure 10.9. Transmission coefficients for varying limestone slab thickness.

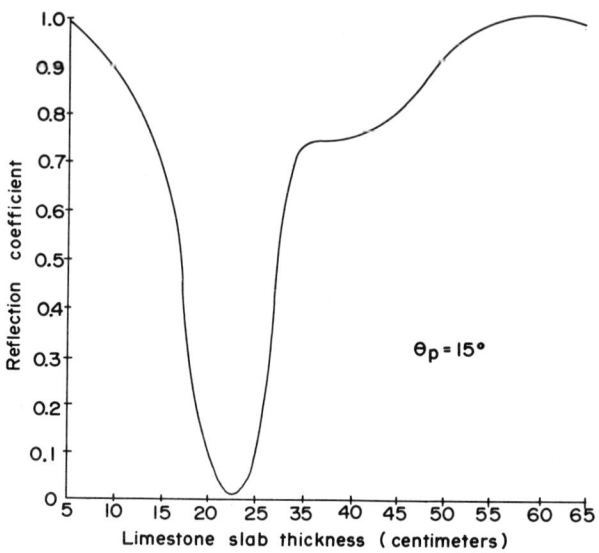

Figure 10.10. Reflection coefficients for varying limestone slab thickness.

angle of the P wave is again taken to be 15°. The transmission coefficient is approximately 1 when the total limestone thickness is 20 cm (two layers), and approximately zero at 53 cm.

All the models above were checked for conservation of energy, and in all cases the requirement of energy conservation is satisfied. In no case was there more than a 0.1 percent discrepancy.

5. SUMMARY

The analysis in this chapter, although quite limited in scope, indicates that the recursive equations can be used to study elastic wave propagation in a one-dimensional inhomogeneous isotropic medium. The models studied were semi-infinite models, and the results are not valid for a medium of finite dimensions in the x-y plane. Much work remains to be done in order to include surface effects, etc. It is too early to predict whether such effects can be included, although an effort is underway to extend the analysis to include diffraction and surface effects.

REFERENCES

1. R. E. Dodd, Propagation of Plane Elastic Waves through an Inhomogeneous Lossless Medium, M.S. Thesis, Vanderbilt University, 1967. Available from University Microfilms, Ann Arbor, Michigan.
2. W. M. Ewing, W. S. Jardetsky, and F. Press, *Elastic Waves in Layered Media*, McGraw-Hill Book Co., New York, 1957.

RELATED REFERENCES

H. Kolsky, *Stress Waves in Solids*, Clarendon Press, Oxford, 1953.
A. E. H. Love, *The Mathematical Theory of Elasticity*, University Press, Cambridge, 1927.
C. B. Officer, *Introduction to the Theory of Sound Transmission*, McGraw-Hill Book Co., New York, 1958.
J. E. White, *Seismic Waves: Radiation, Transmission and Attenuation*, McGraw-Hill Book Co., New York, 1965.

CHAPTER II

Magnon-Phonon Coupling in Ferrites

The interaction of magnon and phonon waves in a ferrite medium is of interest to many workers designing slow wave devices which utilize the interaction between the magnon and phonon waves. Excitation of the magnon or phonon wave at the boundaries of the medium is not discussed. Transducers to excite and detect these waves are an integral part of the design of the devices, but this study would require knowledge of the type of device.

The propagation of magnon and phonon waves in a ferromagnetic crystal was described by Kittel [1]. Kittel considered the total energy in the medium to derive a set of differential equations which describe the interaction process. These differential equations can be reduced to a set of coupled differential equations. Since coupled differential equations are the main topic of this book, the method of handling the coupling problem in ferrites is described. A simple coupling model is described in the following chapter. The ferrite medium is assumed to be homogeneous, and damping of the waves is neglected.

The equations describing the interaction process, as derived by Kittel, are based on a Hamiltonian density, which includes the Zeeman, exchange, magnetoelastic and phonon terms. These equations have been the starting point of much of the analysis that has appeared in journals since that time. The coupled differential equations to be solved are those derived by Sethares [2, 3], who used Kittel's equations to describe the magnon-phonon interaction in ferrimagnetic crystals. Sethares reduced the four equations of Kittel to two coupled equations and added a forcing function to the equations.

The two equations of Sethares are solved in this chapter by using the recursive equations and assuming a continuous medium. The work reported herein has been taken from a thesis submitted by Bradley [4], and the details of the development can be obtained from that work.

I. MAGNON-PHONON EQUATIONS OF MOTION IN FERRITES

Consider a slab of ferrite material of infinite extent in the x-y plane and of finite thickness in the direction of the z axis. The physical system is depicted in Figure 11.1. It is assumed that the ferrite is immersed in a static

11 MAGNON-PHONON COUPLING IN FERRITES

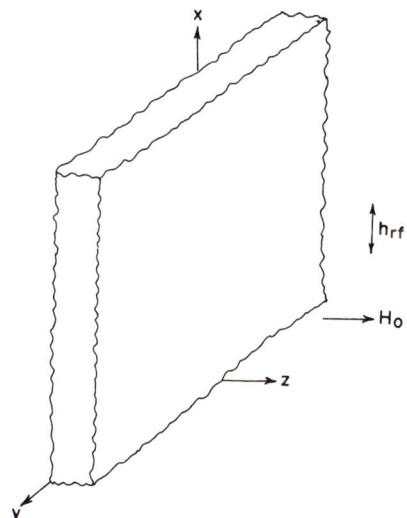

Figure 11.1. Ferrite medium.

magnetic field H_0, which is directed along the z axis. A radio frequency magnetic field of small amplitude is superimposed on the material, and this field is directed along the x axis.

The magnetic system is described by the parameters λ, M, and $\overline{\Delta H}$; λ is the exchange constant, M is the magnetization, and $\overline{\Delta H}$ is the line width of the magnetic material. The acoustic or phonon system has the parameters α, η, and ρ; α is the elastic constant, η is the phonon loss constant, and ρ is the mass density.

The four equations of motion for the small signal component of the magnetization $m(z, t)$ and the crystalline displacement $r(z, t)$ as derived by Sethares, are given below.

$$-\dot{m}_x(z, t) = \gamma H m_y(z, t) - \lambda m_y''(z, t) + \gamma b r_y'(z, t)$$
$$- \gamma M h_y(z, t) + \gamma \overline{\Delta H} m_x(z, t), \qquad (11.1\text{a})$$

$$\dot{m}_y(z, t) = \gamma H m_x(z, t) - \lambda m_x''(z, t) + \gamma b r_x'(z, t)$$
$$- \gamma M h_x(z, t) - \gamma \overline{\Delta H} m_y(z, t), \qquad (11.1\text{b})$$

$$\rho \ddot{r}_x(z, t) + \eta \dot{r}_x(z, t) = \alpha r_x''(z, t) + \frac{b}{M} m_x'(z, t), \qquad (11.1\text{c})$$

$$\rho \ddot{r}_y(z, t) + \eta \dot{r}_y(z, t) = \alpha r_y''(z, t) + \frac{b}{M} m_y'(z, t), \qquad (11.1\text{d})$$

where the parameter b is the magnetoelastic coupling constant, and dots and primes denote time and space derivatives in the usual sense.

The four equations above can be combined in such a way that only two equations need to be considered in the analysis. Let the vector $\psi(z, t)$ be defined as the complex function

$$\psi(z, t) = \psi_x(z, t) + j\psi_y(z, t), \tag{11.2}$$

where $\psi(z, t)$ signifies m, r, or h in (11.1). Using (11.2) in (11.1) and by manipulating the equations, we find

$$\dot{m}(z, t) = (j\gamma H - \gamma \overline{\Delta H})m(z, t) - j\lambda m''(z, t)$$
$$+ j\gamma b r'(z, t) - j\gamma H h(z, t), \tag{11.3a}$$

$$\rho \ddot{r}(z, t) + \eta \dot{r}(z, t) = \alpha r''(z, t) + \frac{b}{M} m'(z, t). \tag{11.3b}$$

It is now assumed that all motion is sinusoidal; that is,

$$\psi(z, t) = \psi(z)e^{j\omega t}. \tag{11.4}$$

The two equations of (11.3) can then be written as

$$\lambda m''(z) + (\omega - \gamma H - j\gamma \overline{\Delta H})m(z) = \gamma b r'(z) - \gamma M h(z), \tag{11.5a}$$

$$\alpha r''(z) + (\rho \omega^2 - j\omega \eta)r(z) = -\frac{b}{M} m'(z). \tag{11.5b}$$

The equations given in (11.5) for $m(z)$ and $r(z)$ are the desired set of equations for the analysis. Since the equations are of second order, the two differential equations are used to generate four first-order ordinary differential equations. Once this has been done, the equations are in the correct form for the recursive equation analysis.

2. THE COUPLED FIRST-ORDER DIFFERENTIAL EQUATIONS

The second-order equations of (11.5) can be reduced to a set of first-order differential equations in the usual way. Let

$$m(z) = u_1(z), \tag{11.6a}$$

$$r(z) = u_2(z), \tag{11.6b}$$

$$m'(z) = v_1(z) = \frac{du_1(z)}{dz}, \tag{11.6c}$$

11 MAGNON-PHONON COUPLING IN FERRITES

and

$$r'(z) = v_2(z) = \frac{du_2(z)}{dz}. \tag{11.6d}$$

If the first equation of (11.5) is divided by λ and the second equation by α, then (11.5) can be written as the four equations to be solved:

$$\frac{du_1(z)}{dz} = v_1(z), \tag{11.7a}$$

$$\frac{du_2(z)}{dz} = v_2(z), \tag{11.7b}$$

$$-\frac{dv_1(z)}{dz} = \left(\frac{\omega - \gamma H - j\gamma \overline{\Delta H}}{\lambda}\right) u_1(z) - \left(\frac{\gamma b}{\lambda}\right) v_2(z) + \left(\frac{\gamma M}{\lambda}\right) h(z), \tag{11.7c}$$

$$-\frac{dv_2(z)}{dz} = \left(\frac{\rho\omega^2 - j\omega\eta}{\alpha}\right) u_2(z) + \left(\frac{b}{\alpha M}\right) v_1(z). \tag{11.7d}$$

The four first-order differential equations can now be written in the matrix form to identify $A(z)$, $B(z)$, $C(z)$, and $D(z)$. These matrices are needed in the analysis because they determine the incremental coefficients. The matrix form of (11.7) is

$$\frac{d}{dz}\begin{bmatrix} u_1(z) \\ u_2(z) \\ -v_1(z) \\ -v_2(z) \end{bmatrix} = \begin{bmatrix} 0 & 0 & 1 & 0 \\ 0 & 0 & 0 & 1 \\ d_{11} & 0 & 0 & c_{12} \\ 0 & d_{22} & c_{21} & 0 \end{bmatrix} \begin{bmatrix} u_1(z) \\ u_2(z) \\ v_1(z) \\ v_2(z) \end{bmatrix} + \begin{bmatrix} 0 \\ 0 \\ f_1(z) \\ 0 \end{bmatrix}, \tag{11.8}$$

where

$$d_{11} = \frac{\omega - \gamma H - j\gamma \overline{\Delta H}}{\lambda}, \tag{11.9a}$$

$$d_{22} = -\frac{\gamma b}{\lambda}, \tag{11.9b}$$

$$c_{12} = \frac{\rho\omega^2 - j\omega\eta}{\alpha}, \tag{11.9c}$$

$$c_{21} = \frac{b}{\alpha M}, \tag{11.9d}$$

and

$$f_1(z) = \frac{\gamma M}{\lambda} h(z). \tag{11.9e}$$

Although the coefficients of (11.9) are considered to be constants, the analysis could be extended to the case in which the coefficients are functions of z. The complete analysis should be carried out, starting from Kittel's equations, when the coefficients are spatially dependent. Care should be taken in deriving the equations since spatial derivatives of the coefficients may occur in the formulation.

The matrix equation (11.8) is now of the correct form and is compatible with (1.1) and (1.2) of Chapter 1:

$$\frac{du(z)}{dz} = B(z)u(z) + A(z)v(z) + e(z), \qquad (1.1)$$

$$-\frac{dv(z)}{dz} = D(z)u(z) + C(z)v(z) + f(z). \qquad (1.2)$$

The functions $u(z)$, $v(z)$, $e(z)$, and $f(z)$ are vectors.

$$u(z) = \begin{bmatrix} u_1(z) \\ u_2(z) \end{bmatrix}, \quad v(z) = \begin{bmatrix} v_1(z) \\ v_2(z) \end{bmatrix}, \quad e(z) = \begin{bmatrix} e_1(z) \\ e_2(z) \end{bmatrix}, \quad f(z) = \begin{bmatrix} f_1(z) \\ f_2(z) \end{bmatrix}.$$

The coefficient matrices $A(z)$, $B(z)$, $C(z)$, and $D(z)$ are given by (11.8) and (11.9).

3. VALUES OF THE CONSTANTS

To carry out the required calculations to determine the interaction, the constants for the magnetic and acoustic systems must be found. A particular ferrite material must be selected, and the constants either found from the literature or determined by measurements. In the analysis which follows, an Mg-Mn ferrite was selected, since this particular ferrite has been studied extensively and data are available for determining the required constants. Many of the constants selected are approximate values, but this does not invalidate the calculations. The constants were obtained from several sources, and there is no assurance that the Mg-Mn ferrites discussed were identical.

The constant γ, the gyromagnetic ratio, is a constant for all ferrites with a value of 1.76×10^7 (sec-oe)$^{-1}$. The saturation magnetization, M or M_s, varies from ferrite to ferrite with the exact composition, and a value of 2000 gauss was picked for this analysis. The line width was assumed to be 175 oersteds.

The exchange constant λ used was calculated from the expression

$$\lambda = \frac{2\gamma A}{M}, \qquad (11.10)$$

11 MAGNON-PHONON COUPLING IN FERRITES

with appropriate values for γ_A and M. For this ferrite, λ was 1.76×10^{-2} (cm²-sec⁻¹). The elastic constant α was chosen as 9.0×10^{11} dynes/cm². The mass density does not vary too greatly from ferrite to ferrite, and an average value of 5.1 gm/cm³ was selected. The mass density depends on the composition of the magnetic material and can be measured precisely.

The phonon loss factor η is related to the ferrite relaxation time T_p through the equation

$$T_p = \frac{2\rho}{\eta}, \tag{11.11}$$

where T_p is somewhat larger than the spin relaxation time of the material. Using a value of spin relaxation of 10^{-8} sec, a value of T_p of 10^{-6} sec was considered to be an approximate value. Then, from (11.11), we obtain a phonon loss factor of 1.02×10^7 gm/cm³-sec.

The last constant needed for the analysis is the magnetoelastic coupling coefficient b. This constant is usually of the order of 10^7 ergs/cm³ and, since an exact value was not available, the value of 10^7 was selected.

The constants given above should not be used in designing a device because they are only approximations. A designer should obtain correct values from the ferrite manufacturer or the design equipment to obtain correct values. It should be kept in mind that the model given here is a very simple one and a device to be built in the laboratory would differ considerably.

4. COMPUTATIONAL RESULTS

The model outlined above is an oversimplified one, but it serves to illustrate the analysis. As has been pointed out earlier, the models selected throughout this book are examined mainly from the computational viewpoint. The analysis of realistic problems is left to the reader.

The analysis which follows considers the wave number to be entirely real with no damping involved. To simulate the anisotropy of the ferrite material, the initial conditions on $m_x(z)$ and $m_y(z)$, as well as $r_x(z)$ and $r_y(z)$ were picked to be unequal. Magnetization components of $m_x(0) = 5.5 \times 10^2$ gauss and $m_y(0) = 4.32 \times 10^2$ gauss were chosen with corresponding phonon amplitudes of $r_x(0) = 1.57 \times 10^{-9}$ cm and $r_y(0) = 1.235 \times 10^{-9}$ cm. The forcing function $f_1(z)$ has a value of 2×10^{12} for the given constants. Since h_{rf} is in quadrature with H_0, the x and y components of h are 1.57×10^{12} and 1.235×10^{12}, respectively.

The calculations to determine $m(z)$ and $r(z)$ and the components thereof were made by using the fundamental matrix equation. Since a forcing function is present, the differential equation is inhomogeneous and the solution is given by

$$\begin{bmatrix} u(x) \\ v(x) \end{bmatrix} = \mathbf{Q}(0, x) \begin{bmatrix} u(0) \\ v(0) \end{bmatrix} + \mathbf{Q}(0, x) \int_0^\infty \mathbf{Q}^{-1}(0, s) \begin{bmatrix} e(s) \\ -f(s) \end{bmatrix} ds;$$

the various functions are defined in (11.7), (11.8), and (11.9). The initial value of the vector $u(0)$ is given above and $v(0)$ was set equal to zero. This is again an assumption which does not hold in a realistic model.

Three different cases were examined by programming the equations given in Section 2 with the constants of Section 3. The three cases are: $m(0) \neq 0$

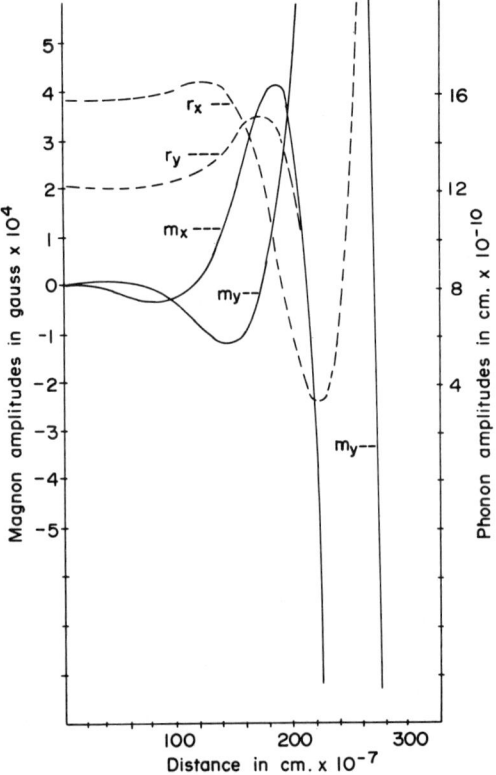

Figure 11.2. Phonon-magnon coupling.

Table 11.1. Magnon and Phonon Amplitudes with $m(0) \neq 0$, $r(0) \neq 0$, and $\Delta z = 1 \times 10^{-7}$ cm

		$z = n \Delta z$		
n	m_x (gauss)	m_y (gauss)	r_x (cm)	r_y (cm)
0	$0.5500E + 03$	$0.4320E + 03$	$0.1570E - 08$	$0.1235E - 08$
20	$0.3513E + 03$	$0.5788E + 03$	$0.1569E - 08$	$0.1233E - 08$
40	$-0.3048E + 03$	$0.8552E + 03$	$0.1569E - 08$	$0.1225E - 08$
60	$-0.1476E + 04$	$0.7597E + 03$	$0.1576E - 08$	$0.1214E - 08$
80	$-0.2866E + 04$	$-0.4859E + 03$	$0.1595E - 08$	$0.1207E - 08$
100	$-0.3352E + 04$	$-0.3581E + 04$	$0.1622E - 08$	$0.1220E - 08$
120	$-0.7303E + 03$	$-0.8242E + 04$	$0.1636E - 08$	$0.1276E - 08$
140	$0.7650E + 04$	$-0.1176E + 05$	$0.1588E - 08$	$0.1378E - 08$
160	$0.2252E + 05$	$0.7737E + 04$	$0.1406E - 08$	$0.1483E - 08$
180	$0.3819E + 05$	$0.1315E + 05$	$0.1041E - 08$	$0.1454E - 08$
200	$0.3735E + 05$	$0.5742E + 05$	$0.5694E - 09$	$0.1062E - 08$
220	$0.9963E + 04$	$0.1157E + 06$	$0.3322E - 09$	$0.7574E - 10$
240	$-0.1340E + 06$	$0.1443E + 06$	$0.1027E - 08$	$-0.1455E - 08$
260	$-0.3298E + 06$	$0.5309E + 05$	$0.3535E - 08$	$0.2738E - 08$
280	$-0.4963E + 06$	$-0.2731E + 06$	$0.8210E - 08$	$0.1792E - 08$
300	$-0.3784E + 06$	$-0.8828E + 06$	$0.1345E - 07$	$0.4390E - 08$

and $r(0) \neq 0$; $m(0) = 0$ and $r(0) \neq 0$; and $m(z) \neq 0$ and $r(0) = 0$. The results, as will be shown, do not differ greatly for the three cases.

The first calculations made were started with $m(0)$ and $r(0)$, as given earlier:

$$m(0) = m_x(0) + jm_y(0) = 5.5 \times 10^2 + j4.32 \times 10^2, \quad (11.12a)$$

$$r(0) = r_x(0) + jr_y(0) = 1.57 \times 10^{-9} + j1.235 \times 10^{-9}; \quad (11.12b)$$

$m'(0)$ and $r'(0)$ were set equal to zero. The calculations were carried out over an interval of 8000 steps with each step, Δz, being equal to 1×10^{-7} cm. The behavior of the magnon and phonon waves for increasing z are shown in Figure 11.2 for this case. As can be seen, the components of $m(z)$, as well as $r(z)$, are not in phase and show different growth patterns. Numerical values for the case of $m(0) \neq 0$ and $r(0) \neq 0$ are tabulated in Table 11.1; values are given for each 100 steps.

The second case studied was one in which a phonon wave was excited at $z = 0$ in the absence of a magnon wave. The calculations were again carried out for 8000 steps, the results being shown in Figure 11.3 for the first 400 steps. The numerical values for this range of z are given in Table 11.2.

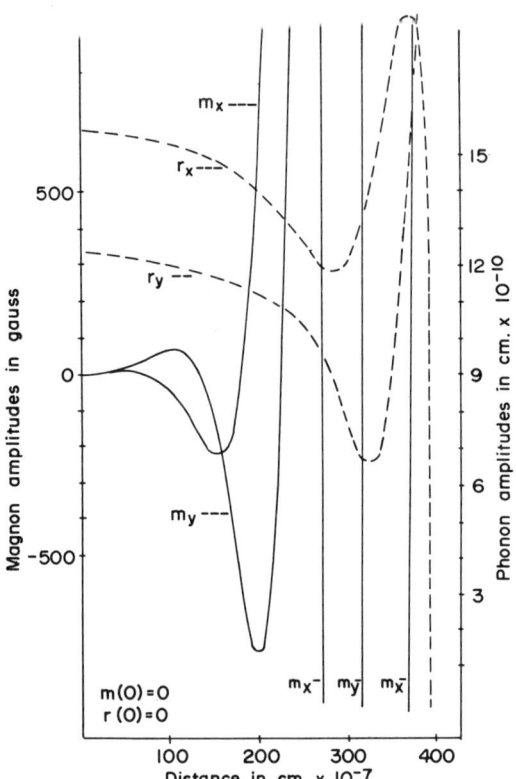

Figure 11.3. Phonon-magnon coupling.

The waves behavior is quite similar to that found when the phonon and magnon waves were both excited at $z = 0$. The major difference noticed is that the growth of the waves is delayed, the first maxima of the phonon and magnon waves occurring at an increased depth within the ferrite. It should be pointed out that the scales on Figures 11.2 and 11.3 are not the same.

The behavior of the system when $m(0) \neq 0$ and $r(0) = 0$ was almost identical with that of the case when $m(0) \neq 0$ and $r(0) \neq 0$. The computed values of $m(z)$ and $r(z)$ agree to three decimal places. The growth of the magnon and phonon waves are thus given by Figure 11.2, and the numerical values are approximately equal to the values given in Table 11.1.

The magnon and phonon waves increase as a function of z as shown in

Table 11.2. Magnon and Phonon Amplitudes with $m(0) = 0$, $r(0) \neq 0$, and $\Delta z = 1 \times 10^{-7}$ cm

		$z = n \Delta z$		
n	m_x (gauss)	m_y (gauss)	r_x (cm)	r_y (cm)
0	0	0	$0.1570E - 08$	$0.1235E - 08$
20	$0.2802E + 01$	$0.2468E + 01$	$0.1568E - 08$	$0.1233E - 08$
40	$0.9207E + 01$	$0.1152E + 02$	$0.1563E - 08$	$0.1229E - 08$
60	$0.1171E + 02$	$0.2968E + 02$	$0.1554E - 08$	$0.1222E - 08$
80	$0.2585E + 01$	$0.5499E + 02$	$0.1542E - 08$	$0.1213E - 08$
100	$-0.4814E + 02$	$0.7265E + 02$	$0.1528E + 02$	$0.1200E - 08$
120	$0.1280E + 03$	$0.4809E + 02$	$0.1510E - 08$	$0.1185E - 08$
140	$-0.2096E + 03$	$0.6821E + 02$	$0.1491E - 08$	$0.1168E - 08$
160	$-0.1979E + 03$	$-0.3088E + 03$	$0.1469E - 08$	$0.1151E - 08$
180	$0.6851E + 02$	$-0.6182E + 03$	$0.1442E - 08$	$0.1134E - 08$
200	$0.7473E + 03$	$0.7536E + 03$	$0.1407E - 08$	$0.1118E - 08$
220	$0.1796E + 04$	$-0.2267E + 03$	$0.1359E - 08$	$0.1098E - 08$
240	$0.2647E + 04$	$0.1574E + 04$	$0.1297E - 08$	$0.1063E - 08$
260	$0.1903E + 04$	$0.4863E + 04$	$0.1230E - 08$	$0.9970E - 09$
280	$-0.2523E + 04$	$0.8462E + 04$	$0.1188E - 08$	$0.8892E - 09$
300	$-0.1218E + 05$	$0.8689E + 04$	$0.1220E - 08$	$0.7504E - 09$

Figures 11.4 and 11.5 for the three cases. The subscripts 1, 2, and 3 identify the waves for cases 1, 2, and 3, respectively. The results are given for the first 500 slabs. Since this is a small signal analysis, the computed values beyond $n = 500$ are of little value other than to show the trend of the curves. It was found that the growth of the magnon and phonon waves continues out to approximately $n = 8000$, after which saturation occurs and there is no amplification after that point.

5. SUMMARY

A simplified model of magnon-phonon interaction in ferrites has been analyzed in this chapter and some of the results presented. The results have been presented in graphical form as well as in tabular form.

Although the model studied has been simplified, there is no reason why the mathematical model cannot be extended to include damping. The extension to a model with finite dimensions in the x-y plane is more difficult and would require considerably more work on the fundamentals of the method. Thus far, in all work, semi-infinite media have been considered.

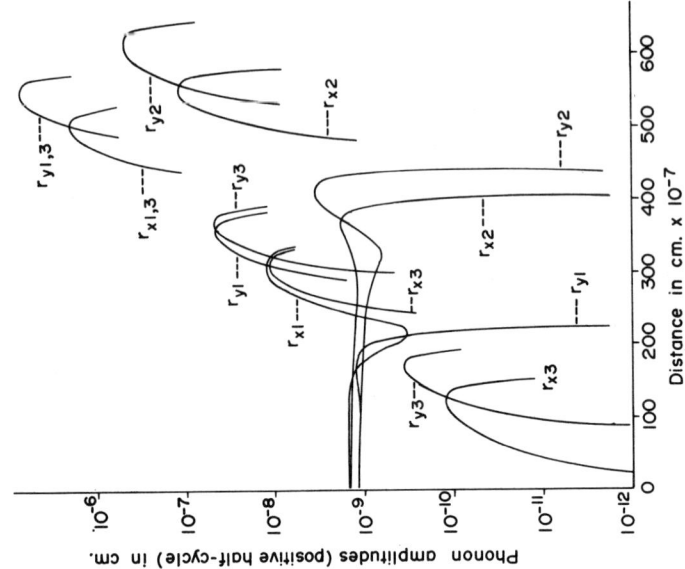

Figure 11.5. Phonon amplitudes vs. distance.

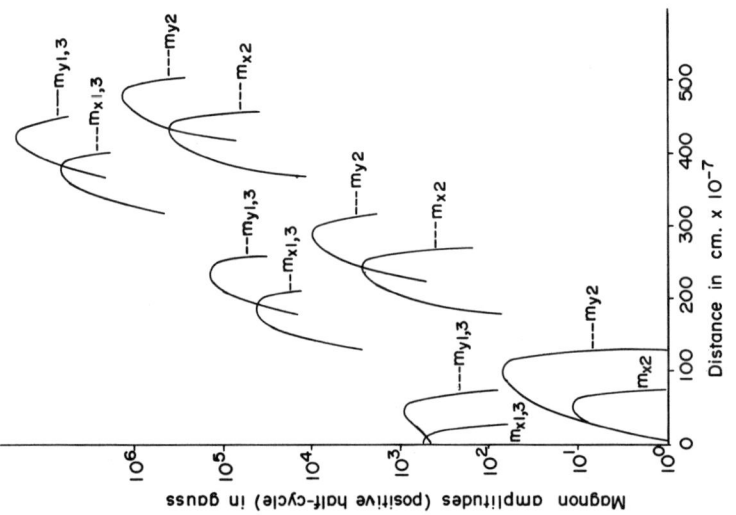

Figure 11.4. Magnon amplitudes vs. distance.

180

A very thorough analysis of the magnon-phonon interaction would be a worthwhile effort. Such a study could possibly lead to better design criteria than do presently exist. The author has not extended the work by Bradley and cannot predict to what extent the invariant imbedding method presented here can be extended to more complicated magnon-phonon interactions.

REFERENCES

1. C. Kittel, Interaction of Spin Waves and Ultrasonic Waves in Ferromagnetic Crystals, *Phys. Rev.*, *110* (1958), 836.
2. J. C. Sethares, Transmission Line Models of Magnon-Phonon Modes in Ferrites: Part I. Uncoupled Modes, Air Force Cambridge Research Laboratories, Rept. No. AFCRL-66-633(1), 1966.
3. J. C. Sethares, Transmission Line Models of Magnon-Phonon Modes in Ferrites: Part II. Coupled Modes, Air Force Cambridge Research Laboratories, Rept. No. AFCRL-67-0314, 1967.
4. W. T. Bradley, Jr., An Invariant Imbedding Solution of Coupled Phonon-Magnon Modes in Ferrites, M.S. Thesis, Vanderbilt University, 1968. Available from University Microfilms, Ann Arbor, Michigan.

CHAPTER 12

Polarization of an Electromagnetic Wave in an Inhomogeneous Ionized Medium

The propagation of an electromagnetic wave in an ionized medium has been studied by many workers. The determination of the wave properties in such a medium is a difficult task even for the simplest inhomogeneous distribution that can be mathematically described. Models of inhomogeneous ionized media that are assumed in order to find an analytical solution are, in most cases, unrealistic. Such distributions of electrons in an ionized medium are seldom found in nature. The determination of the behavior of an electromagnetic wave in an inhomogeneous medium using numerical analysis is described in this chapter.†

It would require an entire book rather than a single chapter to describe completely the behavior of a wave in an inhomogeneous ionized medium. The work discussed in this chapter is limited to a small part of the overall problem, in particular, to the coupling of energy between the ordinary and extraordinary waves. An inhomogeneous ionized layer immersed in a static magnetic field is considered. The coupled mode numerical method is used for the analysis. The ionized layer is assumed to have an electron density $N(z)$ of an Epstein profile. This distribution is not a realistic model for the ionosphere, but there is no reason why the method cannot be extended to any desired electron distribution. A static magnetic field is present everywhere, and it is assumed that the propagation direction of the wave and the direction of the magnetic field do not coincide.

It is not difficult to show that an electromagnetic wave in an ionized medium immersed in a magnetic field must break into two waves propagating in each direction. These waves have different wave numbers and are referred to as the ordinary and extraordinary waves. This chapter attempts to analyze the coupling between these waves under the conditions above. Ratcliffe [2] and Budden [3] have discussed in detail the problem posed here. In either of these

---

† The material in this chapter has been condensed from the dissertation of Hugh A. Davies [1], carried out under the author's supervision.

12 POLARIZATION OF AN ELECTROMAGNETIC WAVE

references, the reader will find the details which cannot be presented here. Space does not permit the inclusion of all of the work in this book.

I. THE INHOMOGENEOUS IONIZED MEDIUM

The ionized layer is considered to be semi-infinite in the x-y plane, the free electron density varying spatially in the direction of z. Let $N(z)$ denote the free electron density. The static magnetic field B_0 has an assumed direction and magnitude; it is therefore a vector.

The refractive index for a medium containing free electrons and with a static magnetic field present is given by

$$\eta^2 = 1 - \frac{X}{1 - jZ - \frac{Y_T^2}{2(1 - X - jZ)} \pm \frac{1}{2}\left[\frac{Y_T^4}{(1 - X - jZ)^2} + 4Y_L^2\right]^{1/2}}. \quad (12.1)$$

This equation is the Appleton-Hartree formula. It is assumed that X and Z are spatially varying quantities, Y_L and Y_T are the longitudinal and transverse components of Y with respect to the propagating wave direction. The function $X(z)$ is given by

$$X(z) = \frac{N(z)e^2}{\varepsilon_0 m \omega^2}, \quad (12.2)$$

Y is defined as a vector,

$$\bar{Y} = \frac{e}{m\omega} \bar{B}; \quad (12.3)$$

and $Z(z)$ is directly proportional to the electron collision frequency,

$$Z(z) = \frac{\nu(z)}{\omega}. \quad (12.4)$$

The free electron density is $N(z)$, e is the charge on an electron, ε_0 is the permittivity of free space, m is the mass of an electron, ω is the angular frequency of the electromagnetic wave, and $\nu(z)$ is a measure of the number of collisions an electron makes per unit time. It is assumed that $\nu(z)$ is a constant such that Z is small.

It is well known that a plane wave is entirely reflected at a point in a medium where the refractive index becomes zero. Since both ordinary and extraordinary waves propagate in an ionized medium, the square of the refractive index as given by (12.1) takes on two values; let these values be denoted by O and XO. Typical plots of η^2 for $Y < 1$ and $Y > 1$ are shown in Figures 12.1 and 12.2, respectively.

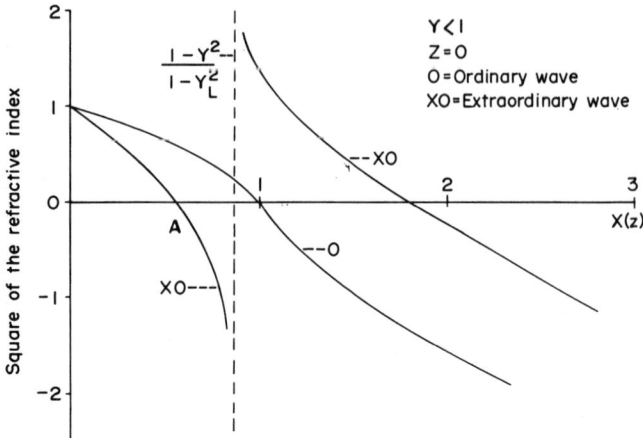

Figure 12.1. Square of the refractive index vs. $X(z)$.

If we assume that a wave is completely reflected when $\eta^2 = 0$, the ordinary (O) wave and the extraordinary (XO) waves undergo complete reflection for $0 < x \leq 1$, when $Y < 1$. It is possible that complete reflection may not occur; some of the energy of the wave which undergoes "complete reflection" may be transferred to the other wave. For example, η^2 for the XO wave equals zero at A, yet the value of η^2 for the O wave is greater than zero. Whether

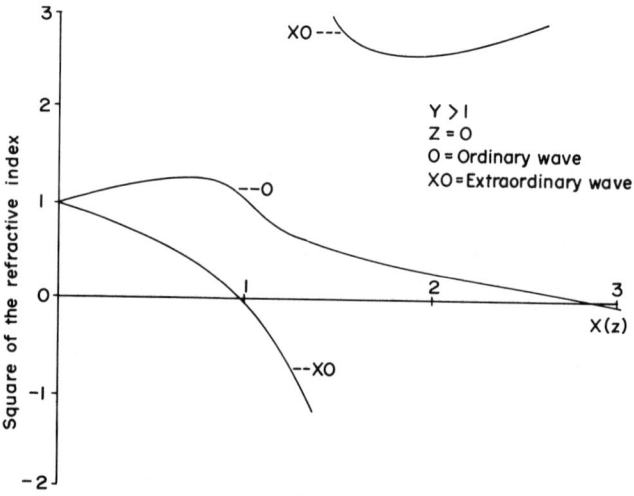

Figure 12.2. Square of the refractive index vs. $X(z)$.

12 POLARIZATION OF AN ELECTROMAGNETIC WAVE

coupling occurs or not is not revealed by an examination of the η^2 curves for the ordinary and extraordinary waves. The wave equations for the two waves in an inhomogeneous medium must be examined to determine if coupling takes place.

The coupled mode equations for the spatial components of the electric field are derived and numerical solutions obtained for several sample cases. The ionized medium which was examined has an electron density distribution described by the Epstein profile. The electron density $N(z)$ can be fixed for the purpose of examining the coupling between the waves. Since $X(z)$ is a frequency-dependent function, different portions of the curves shown in Figures 12.1 and 12.2 can be examined by changing the frequency.

The Epstein profile specifies an electron density distribution as given by the equation

$$N(z) = \frac{4M \exp[m(z - z_0)]}{\{1 + \exp[m(z - z_0)]\}^2}. \tag{12.5}$$

The magnitude factor M was selected for a maximum density which occurs at the point $z = z_0$. The distribution is symmetrical about z_0. Since $N(z)$ is finite for all z, the Epstein layer was terminated at a thickness L_0. The electron density at the end points had a finite discontinuity which was taken into account in the calculations.

The distribution above is not in the best form to start the calculations at $z = 0$, which was preferred. Equation (12.5) can be modified so that $z = 0$ specifies the left boundary of the layer and $z = L_0 = 1$ the right boundary, as shown in Figure 12.3. If $L_0 = 2z_0$ is set equal to unity, then the profile has a

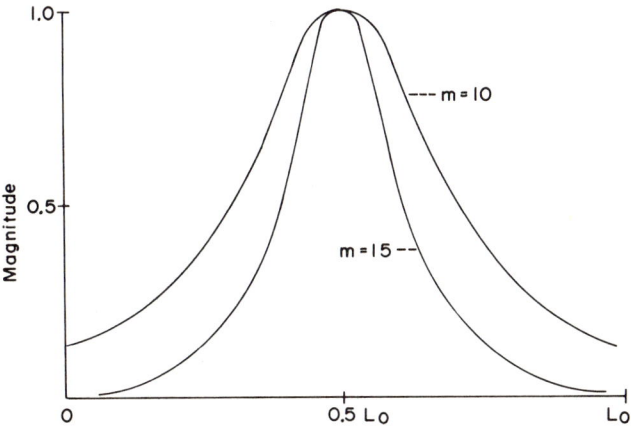

Figure 12.3. Epstein profile.

Table 12.1. Constants for Sample Problems

Freq. (mHz)	N_{max} (Elec./cm³)	X_{max} (Dimensionless)	$|\bar{Y}|$ (Dimensionless)	k_0 ($\frac{1}{10}$ meter)
3.16	1.86×10^5	1.50	0.443	0.600
1.0	3.10×10^4	2.50	1.400	0.209

$Z = 5.0 \times 10^{-5}$, $\quad |\bar{B}| = 5 \times 10^{-5}$ weber, $\quad m = 30$

normalized length of L_0. The width of the barrier can be modified by varying the propagation coefficient k_0. The wave number k_0 is the only constant which has a dimension; all the other constants are dimensionless. The constants of the Epstein profile used in the calculations are shown in Table 12.1.

The electron density for a frequency of 1 mHz was taken to be lower than that for the other cases. The value of X_{max} for the higher electron densities leads to numerical difficulties in calculating the incremental coefficients for the recursive equations. A change in the electron density rather than a change in the entire computer program was the simplest way to circumvent the difficulty.

2. THE COUPLED MODE EQUATIONS

Assume that a plane electromagnetic wave is incident on an ionized layer, as shown in Figure 12.4. The incident wave is taken as linearly polarized, and the wave vector is constrained to lie in the x-z plane. The magnetic field is fixed in the x-z plane and two angles, 30° and 60° measured from the x axis, are selected for the study.

If the magnetic field of the wave, H, is normalized to Z_0, $H = Z_0 \hat{H}$, where $\hat{H}$ is the unnormalized field, then the field equations for the problem can be written from Maxwell's equations:

$$\frac{\partial E_z}{\partial x} - \frac{\partial E_y}{\partial z} = -jk_0 H_x, \qquad \frac{\partial H_z}{\partial y} - \frac{\partial H_y}{\partial z} = j\frac{k_0}{\varepsilon_0} D_x,$$

$$\frac{\partial E_x}{\partial z} - \frac{\partial E_z}{\partial x} = -jk_0 H_y, \qquad \frac{\partial H_x}{\partial z} - \frac{\partial H_z}{\partial x} = j\frac{k_0}{\varepsilon_0} D_y, \qquad (12.6)$$

$$\frac{\partial E_y}{\partial x} - \frac{\partial E_x}{\partial y} = -jk_0 H_z, \qquad \frac{\partial H_y}{\partial x} - \frac{\partial H_x}{\partial y} = j\frac{k_0}{\varepsilon_0} D_z.$$

12 POLARIZATION OF AN ELECTROMAGNETIC WAVE

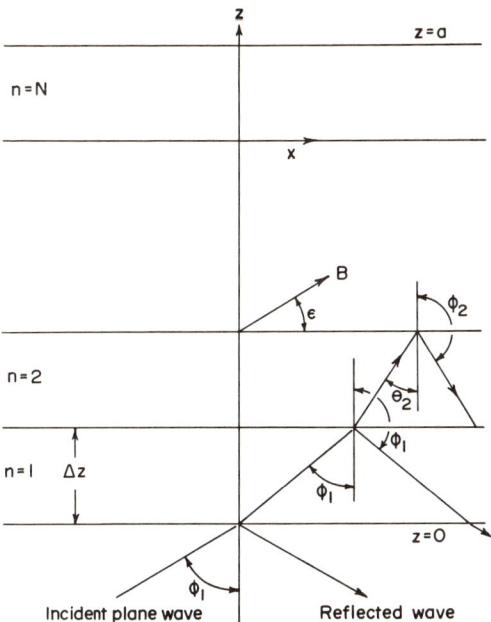

Figure 12.4. Reflection process for ionized layer.

The propagation coefficient for free space, k_0, is defined by

$$k_0 = \frac{2\pi}{\lambda_0} = \frac{\omega}{c_0} = \omega\sqrt{\varepsilon_0\mu_0}. \tag{12.7}$$

It is not difficult to show by use of Figure 12.4 that Snell's law requires that

$$\eta_n \sin \theta_n = \sin \theta_I, \tag{12.8}$$

where η_n is the refractive index for the nth layer. Since the propagating fields are assumed to be of the plane-wave type, the field in each layer can be represented mathematically by the equation

$$A_n \exp\{-jk_0\eta_n(z \cos \theta_n + x \sin \theta_n)\}$$
$$+ B_n \exp\{jk_0\eta_n(z \cos \theta_n - x \sin \theta_n)\}, \tag{12.9}$$

where A_n and B_n are the amplitudes of the positive and negative going waves in the medium. If the operators $\partial/\partial x$ and $\partial/\partial y$ are applied to (12.9), we find from (12.8) and (12.9) that the operators are mathematically equivalent to multiplication; that is,

$$\frac{\partial}{\partial x} = jk_0 S, \tag{12.10a}$$

$$\frac{\partial}{\partial y} = 0, \tag{12.10b}$$

where $S \triangleq \sin \theta_I$ and, for later use, $C \triangleq \cos \theta_I$.

Equations (12.10a) and (12.10b) can now be used to eliminate the partial derivatives with respect to x and y.

$$\frac{dE_y}{dz} = jk_0 H_x, \tag{12.11a}$$

$$\frac{dE_x}{dz} = -jk_0 SE_z - jk_0 H_y, \tag{12.11b}$$

$$-jk_0 SE_y = -jk_0 H_z, \tag{12.11c}$$

$$\frac{dH_y}{dz} = -j\frac{k_0}{\varepsilon_0} D_x, \tag{12.11d}$$

$$\frac{dH_x}{dz} = j\frac{k_0}{\varepsilon_0} D_y - jk_0 SH_z, \tag{12.11e}$$

$$-jk_0 SH_y = j\frac{k_0}{\varepsilon_0} D_z. \tag{12.11f}$$

The equations above are not in the desired forms since components of $\bar{D}$, $\bar{E}$, $\bar{H}$ are involved. The dependence on $\bar{D}$ can be eliminated by making use of

$$\bar{D} = \varepsilon_0 \bar{E} + \bar{P}, \tag{12.12}$$

where $\bar{P}$ is defined as

$$\begin{bmatrix} P_x \\ P_y \\ P_z \end{bmatrix} = \frac{\varepsilon_0 X}{U(U^2 - Y^2)} \times \begin{bmatrix} U^2 - \alpha^2 Y & -j\gamma YU - \alpha\beta Y^2 & j\beta YU - \alpha\gamma Y^2 \\ j\gamma YU - \alpha\beta Y^2 & U^2 - \beta^2 Y^2 & -j\alpha YU - \beta\gamma Y^2 \\ -j\beta YU - \alpha\gamma Y^2 & j\alpha YU - \beta\gamma Y^2 & U^2 - \gamma^2 Y^2 \end{bmatrix} \begin{bmatrix} E_x \\ E_y \\ E_z \end{bmatrix}. \tag{12.13}$$

The term U in (12.13) is defined as $(1 - jZ)$, and α, β, and γ are the direction cosines of $\bar{Y}$. The 3×3 matrix of (12.13) is the electric susceptibility M

12 POLARIZATION OF AN ELECTROMAGNETIC WAVE

whose elements are denoted by M_{ij}, where i and j take on the values x, y, and z.

By using (12.12) and (12.13), the eight equations of (12.11) can be reduced to four differential equations defining H_x, H_y, E_x, and E_y:

$$\frac{dE_x}{dz} = \frac{jk_0}{1 + M_{zz}} \{SM_{zx}E_x + SM_{zy}E_y - (C^3 + M_{zz})H_y\}, \quad (12.14a)$$

$$\frac{dE_y}{dz} = jk_0 H_x, \quad (12.14b)$$

$$\frac{dH_x}{dz} = \frac{jk_0}{1 + M_{zz}} \{[M_{yx}(1 + M_{zz}) - M_{yz}M_{zx}]E_x$$
$$+ [(C^2 + M_{yy})(1 + M_{zz}) - M_{yz}M_{zy}]E_y - SM_{yz}H_y\}, \quad (12.14c)$$

$$\frac{dH_y}{dz} = \frac{jk_0}{1 + M_{zz}} \{[-(1 + M_{zz})^2 + M_{xz}M_{zx}]E_x$$
$$- [M_{xy}(1 + M_{zz}) - M_{xz}M_{zy}]E_y + SM_{xz}H_y\}. \quad (12.14d)$$

The equations given in (12.14) are in coupled mode form. The analysis of this set determines E_x, E_y, H_x, and H_y throughout the ionized medium. Rather than solve for these four variables, H_x and H_y were eliminated from (12.14) such that a set of second-order differential equations in terms of E_x and E_y were found.

$$\frac{d^2E_x}{dz^2} = -C_{11}\frac{dE_x}{dz} - C_{12}\frac{dE_y}{dz} - D_{11}E_x - D_{12}E_y, \quad (12.15a)$$

$$\frac{d^2E_y}{dz^2} = -C_{21}\frac{dE_x}{dz} - D_{21}E_x - D_{22}E_y. \quad (12.15b)$$

The coefficients given in (12.15) are found from (12.14).

$$C_{11} = \frac{-1}{1 + M_{zz}}\left[jk_0 SM_{zx} + jk_0 SM_{xz} + \frac{S}{C^2 + M_{zz}}\left(\frac{dM_{zz}}{dz}\right)\right], \quad (12.16a)$$

$$C_{12} = -jk_0 \frac{SM_{zy}}{1 + M_{zz}}, \quad (12.16b)$$

$$C_{21} = -jk_0 \frac{SM_{yz}}{1 + M_{zz}}, \quad (12.16c)$$

$$C_{22} = 0, \quad (12.16d)$$

$$D_{11} = \frac{1}{(1+M_{zz})^2}\left\{jk_0S\left[\left(\frac{dM_{zx}}{dz}\right)(1+M_{zz}) - M_{zx}\left(\frac{dM_{zz}}{dx}\right)\right]\right.$$
$$- k_0^2(C^2+M_{zz})[(1+M_{zz})(1+M_{xx}) - M_{xz}M_{zx}]$$
$$\left. + k_0^2S^2M_{zx}M_{xz} - jk_0SM_{zx}\frac{S^2}{C^2+M_{zz}}\left(\frac{dM_{zz}}{dz}\right)\right\}, \qquad (12.16e)$$

$$D_{12} = -\frac{1}{(1+M_{zz})^2}\left[S(1+M_{zz})\left(\frac{dM_{zy}}{dz}\right) - SM_{zy}\left(\frac{dM_{zz}}{dz}\right)\right.$$
$$- k_0^2(C^2+M_{zz})[M_{xy}(1+M_{zz}) - M_{xy}M_{zy}]$$
$$\left. + k_0^2S^2M_{zy}M_{xz} - jk_0SM_{zy}\frac{S^2}{C^2+M_{zz}}\left(\frac{dM_{zz}}{dz}\right)\right], \qquad (12.16f)$$

$$D_{21} = \frac{k_0^2}{1+M_{zz}}\left[M_{yx}(1+M_{zz}) - M_{yz}M_{zx} - \frac{S^2M_{zx}M_{yz}}{C^2+M_{zz}}\right], \qquad (12.16g)$$

$$D_{22} = \frac{k_0^2}{1+M_{zz}}\left[(C^2+M_{yy})(1+M_{zz}) - M_{yz}M_{zy} - \frac{S^2M_{zy}M_{yz}}{C^2+M_{zz}}\right]. \qquad (12.16h)$$

The equations given in (12.15) and the coefficients of (12.16) define the coupled mode differential equations which will be solved. The four equations of the coupled mode problem are given in matrix form in (12.17), where u_1 and u_2 are the electric field components E_x and E_y. The derivatives of the field components are given by $v_1(x)$ and $v_2(x)$.

$$\begin{bmatrix} \dfrac{du_1}{dz} \\ \dfrac{du_2}{dz} \\ -\dfrac{dv_1}{dz} \\ -\dfrac{dv_2}{dz} \end{bmatrix} = \begin{bmatrix} 0 & 0 & 1 & 0 \\ 0 & 0 & 0 & 1 \\ D_{11} & D_{12} & C_{11} & C_{12} \\ D_{21} & D_{22} & C_{21}' & C_{22} \end{bmatrix} \begin{bmatrix} u_1 \\ u_2 \\ v_1 \\ v_2 \end{bmatrix}. \qquad (12.17)$$

The matrix equation of (12.17) defines the coupled mode problem which must be solved. Equation (12.17) is in the most general form. To study the coupling between the ordinary and extraordinary wave, a means of breaking the general solution into the ordinary and extraordinary waves must be available.

3. THE BOOKER QUARTIC, POLARIZATION, AND THE DIRECTION OF PROPAGATION

To determine the polarization of the waves and the direction of propagation, the first-order equations of (12.14) must be examined. Assume that E_x, E_y, H_x, and H_y must have solutions of the form

$$\psi(z) = \tilde{\psi}_0 e^{-jk_0 qz}, \tag{12.18}$$

where $\psi(z)$ denotes the four field components and the tilde ($\sim$) indicates the amplitude constant. The operator d/dz which appears in (12.14) can then be written as

$$-\frac{1}{jk_0}\frac{d}{dz} = q. \tag{12.19}$$

In order that there be a solution to (12.14), the determinant of the matrix

$$\begin{bmatrix} -\dfrac{SM_{zx}}{1+M_{zz}} - q & \dfrac{-SM_{zy}}{1+M_{zz}} & 0 & \dfrac{C^2+M_{zz}}{1+M_{zz}} \\ 0 & -q & -1 & 0 \\ -M_{yx} + \dfrac{M_{yz}M_{zx}}{1+M_{zz}} & -C^2 - M_{yy} + \dfrac{M_{zy}M_{yz}}{1+M_{zz}} & -q & \dfrac{SM_{yz}}{1+M_{zz}} \\ 1+M_{xx} - \dfrac{M_{xz}M_{zz}}{1+M_{zz}} & M_{xy} - \dfrac{M_{xz}M_{zy}}{1+M_{zz}} & 0 & -\dfrac{SM_{xz}}{1+M_{zz}} - q \end{bmatrix}$$

(12.20)

must be zero. The determinant of (12.20) can be written as the quartic equation commonly called the Booker quartic:

$$a_4 q^4 + a_3 q^3 + a_2 q^2 + a_1 q + a_0 = 0. \tag{12.21}$$

The coefficients a_4, a_3, a_2, a_1, and a_0 can be obtained from (12.13) and (12.14).

$$a_4 = U(U^2 - Y^2) + X(\gamma^2 Y^2 - U^2), \tag{12.22a}$$

$$a_3 = 2\gamma X Y^2 S, \tag{12.22b}$$

$$a_2 = -2U(U-X)(C^2 U - X) + 2Y^2(C^2 U - X) + XY^2(1 - C^2\alpha^2 + \alpha^2 S^2), \tag{12.22c}$$

$$a_1 = -2C^2 \gamma X Y^2 S, \tag{12.22d}$$

$$a_0 = (U-X)(C^2 U - X)^2 - C^2 Y^2(C^2 U - X) - C^2 X Y^2 S^2 \alpha^2. \tag{12.22e}$$

The four roots of the quartic given in (12.21) correspond to two waves in the direction of increasing z and two waves for decreasing z. These roots, in general, are complex and the sign of the imaginary part of q determines the direction of energy flow. If $\text{Im}(q)$ is negative, the energy flow is in the direction of positive z; positive $\text{Im}(q)$ signifies energy flow in the negative z direction.

Since the characteristic equation of (12.21) is a quartic and has four roots, there are four component waves at each point in the medium. The determination of these four components is essential to the polarization analysis. Assume that each of these four waves has a propagation coefficient $k = k_0 q_i$, where $i = 1, 2, 3, 4$, and let E_x and E_y be constructed as a linear combination of the four waves:

$$E_x = E_{x1} + E_{x2} + E_{x3} + E_{x4}, \tag{12.23a}$$

$$E_y = E_{y1} + E_{y2} + E_{y3} + E_{y4}, \tag{12.23b}$$

where $E_{xi} = \tilde{E}_{xi} \exp(-jk_0 q_i z)$, with a similar expression for y. If ρ_i is the wave polarization for each wave associated with a root of the quartic, q_i, we find that ρ_i is given by

$$\rho_i = \frac{E_{yi}}{E_{xi}} = \frac{A - JD/(q_i - G)}{-q_i^2 - B - [JF/(q_i - G)]}, \quad i = 1, \ldots, 4, \tag{12.24}$$

where

$$A = -M_{yx} + \frac{M_{yz} M_{zy}}{1 + M_{zz}}, \tag{12.25a}$$

$$B = -C^2 - M_{yy} + \frac{M_{yz} M_{zy}}{1 + M_{zz}}, \tag{12.25b}$$

$$D = 1 + M_{xx} - \frac{M_{xz} M_{zx}}{1 + M_{zz}}, \tag{12.25c}$$

$$F = M_{xy} - \frac{M_{xz} M_{zy}}{1 + M_{zz}}, \tag{12.25d}$$

$$G = -\frac{SM_{xz}}{1 + M_{zz}}, \tag{12.25e}$$

$$J = \frac{SM_{yz}}{1 + M_{zz}}. \tag{12.25f}$$

It is not difficult to show that, if E_x, E_y, dE_x/dx, dE_y/dy, q_i, and ρ_i are known, the four components of the waves can be found. To find the components of E_x and E_y as given in (12.23), differentiate the two expressions of

12 POLARIZATION OF AN ELECTROMAGNETIC WAVE

(12.23) and eliminate the components of E_y by using (12.24). Four equations for the four unknowns, E_{xi}, can then be found. These simultaneous equations can be solved to determine the field components. Knowing E_{xi} for all values of z makes it possible to find E_{yi} by using (12.24) where it is assumed that ρ_i is known everywhere.

4. DETERMINATION OF THE BOUNDARY CONDITIONS

The four coupled equations given in (12.17) can be solved to determine E_x and E_y everywhere, provided that initial conditions or boundary-values are specified. It has been shown in a preceding chapter that the transmission or the fundamental matrix can be calculated independently of the boundary conditions for a linear set of coupled differential equations. The functions E_x and E_y can then be calculated provided that the proper boundary conditions have been specified. The propagation of an electromagnetic wave through an ionized medium does not lead to a simple boundary condition. To appreciate the difficulty in specifying the boundary conditions, consider Figure 12.4. It can be assumed that a wave is incident on the medium at $z = 0$, and the magnitude of this wave can be taken as unity. Such a specification does not indicate the values of E_x and E_y at that point since these components are made up of an incident and a reflected wave.

The known boundary condition which can be imposed at $z = a$ is that there is no incident wave propagating in the direction of negative z. Since it is assumed that the medium for $z > a$ is free space, there is no reflected wave. This boundary condition, along with the arbitrary initial condition, completes the specification of the boundary conditions necessary to solve the problem.

To see how the conditions above are used, consider the two equations of (12.15) for values of $z < 0$ and $z > a$. The wave equations for these two regions are

$$\frac{d^2 E_x}{dz^2} + k^2 E_x = 0, \qquad (12.26\text{a})$$

$$\frac{d^2 E_y}{dz^2} + k^2 E_y = 0, \qquad (12.26\text{b})$$

where $k^2 = k_0^2 C^2$. Equations (12.26a) and (12.26b) can be written compactly as a matrix equation,

$$\frac{d^2 \psi(z)}{dz^2} + \mathbf{K}^2 \psi(z) = 0, \qquad (12.27)$$

where $\psi(z)$ is the column vector

$$\psi(z) = \begin{vmatrix} E_x(z) \\ E_y(z) \end{vmatrix} \qquad (12.28)$$

and $\mathbf{K}^2 = k^2[I]$. The general plane wave solution to (12.27) can be written as

$$\psi(z) = \psi_+(z) + \psi_-(z), \qquad (12.29)$$

where the plus sign indicates the wave in the positive z direction and minus sign the wave in the opposite direction. Assume that waves traveling in the positive z direction vary as $\exp(-j\mathbf{K}z)$, whereas those in the negative direction vary as $\exp(j\mathbf{K}z)$. Using the assumed solution of (12.29) and the derivative of $\psi(z)$, it is easy to show that

$$\psi_+(z) = \tfrac{1}{2}\psi(z) + \frac{j}{2}\mathbf{K}^{-1}\psi'(z), \qquad (12.30a)$$

$$\psi_-(z) = \tfrac{1}{2}\psi(z) - \frac{j}{2}\mathbf{K}^{-1}\psi'(z), \qquad (12.30b)$$

where the prime (') indicates differentiation with respect to z.

The wave functions $\psi(a)$ and $\psi(0)$, as well as their derivatives, must satisfy the matrix equation of (12.31).

$$\begin{bmatrix} \psi(a) \\ \psi'(a) \end{bmatrix} = \begin{bmatrix} Q_{11}(0, a) & Q_{12}(0, a) \\ Q_{21}(0, a) & Q_{22}(0, a) \end{bmatrix} \begin{bmatrix} \psi(0) \\ \psi'(0) \end{bmatrix}. \qquad (12.31)$$

Substituting $\psi(a)$ and $\psi'(a)$ from (12.31) into (12.30b), we find at $z = a$

$$\psi_+(a) = \tfrac{1}{2}\{Q_{11}(0, a)\psi(0) + Q_{12}(0, a)\psi'(0)\}$$
$$+ \frac{j}{2}\mathbf{K}^{-1}\{Q_{21}(0, a)\psi(0) + Q_{22}(0, a)\psi'(0)\}. \qquad (12.32)$$

This is the boundary condition for the wave in the direction of positive z. The negative going wave at $z = 0$ is given by

$$\psi_-(0) = \tfrac{1}{2}\psi(0) - \frac{j}{2}\mathbf{K}^{-1}\psi'(0). \qquad (12.33)$$

Equations (12.32) and (12.33) can now be used to find $\psi(0)$ and $\psi'(0)$ in terms of the available initial and boundary conditions. Since $\psi_+(0)$ and $\psi_-(a)$ are specified, $\psi(0)$ and $\psi'(0)$ are known. The initial conditions for $\psi(0)$ and

12 POLARIZATION OF AN ELECTROMAGNETIC WAVE

$\psi'(0)$ are

$$\psi(0) = \left[\frac{\beta \mathbf{K} - j\alpha}{2}\right]^{-1}\left[\beta \mathbf{K}\psi_+(0) - \frac{j}{2}\psi_-(a)\right], \tag{12.34a}$$

$$\psi'(0) = \beta^{-1}\psi_-(a) - \beta^{-1}\alpha\psi(0), \tag{12.34b}$$

where

$$\alpha = \tfrac{1}{2}Q_{11}(0, a) - \frac{j}{2}\mathbf{K}^{-1}Q_{21}(0, a), \tag{12.34c}$$

$$\beta = \tfrac{1}{2}Q_{12}(0, a) - \frac{j}{2}\mathbf{K}^{-1}Q_{22}(0, a). \tag{12.34d}$$

The column vectors $\psi(0)$ and $\psi'(0)$ constitute the needed boundary conditions at the point $z = 0$.

5. SOME NUMERICAL RESULTS

The coupling between the ordinary and extraordinary waves in the ionized medium defined in terms of the Epstein profile was obtained numerically by solving the equations above. Three frequencies, 1.0 mHz, 3.16 mHz, and 10 mHz, were chosen for the calculations, although data for only the first two frequencies are given here. The physical constants for the Epstein layer are given in Table 12.1. The magnetic field direction was selected to be 30° and 60° from the x axis with the vector lying in the x-z plane, as shown in Figure 12.4.

The angle of the incident wave was varied from 0° to 75°, the particular values of 0°, 5°, 10°, 22°, 27°, 28°, 29°, 30°, 32°, 45°, 55°, 58° 59°, 60°, 62°, and 75° being used for each frequency and magnetic field. Small variations in the incident angle near 30° were made to study the polarization change which occurred very abruptly near these values when the magnetic field direction was 30°. The same abrupt change occurred at 60° when the direction of the magnetic field was 60°.

The coupling between the ordinary and extraordinary waves is illustrated in Figures 12.5, 12.6, and 12.7. These figures are plots of R_s vs. θ, where R_s is defined as the ratio of the time-averaged power flow per unit area for the wave in the negative z direction to that of the wave in the positive z direction.

$$R_s = \frac{\hat{P}_-(\text{unit area})}{\hat{P}_+(\text{unit area})}. \tag{12.35}$$

A few conclusions can be drawn from the data presented in the three

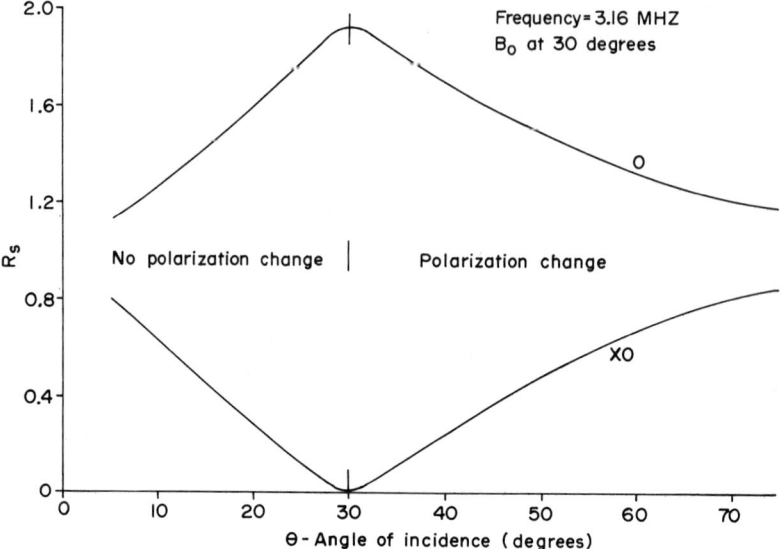

Figure 12.5. R_s vs. angle of incidence.

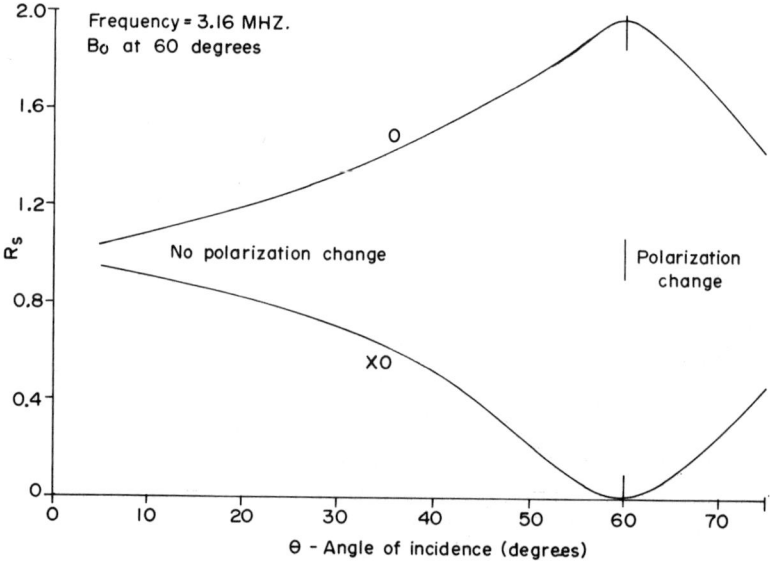

Figure 12.6. R_s vs. angle of incidence.

12 POLARIZATION OF AN ELECTROMAGNETIC WAVE

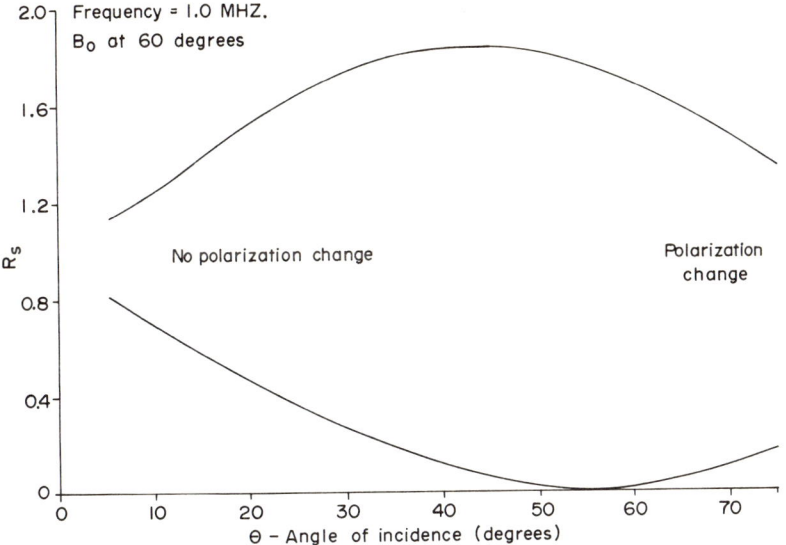

Figure 12.7. R_s vs. angle of incidence.

figures. We observe first that there is a polarization change when the directions of the magnetic field and of the incident wave coincide. Secondly, we see that R_s for the ordinary wave is a maximum when the incident angle is approximately equal to the angle of the magnetic field, whereas R_s for the extraordinary wave is a minimum. This, perhaps, indicates that maximum power transfer occurs between the ordinary and extraordinary waves under these conditions.

To illustrate the coupling in a different way, let P^{xo} and P^o be the power in the extraordinary and ordinary waves, respectively, then define

$$R_s^o = \frac{\hat{P}_-^o + \delta}{\hat{P}_+^o}, \tag{12.36a}$$

$$R_s^{xo} = \frac{\hat{P}_-^{xo} - \delta}{\hat{P}_+^{xo}}, \tag{12.36b}$$

where δ is defined as the net power transfer. Solving for δ, we find

$$\delta = \hat{P}_+^o R_s - \hat{P}_-^o. \tag{12.37}$$

If $\hat{P}_+^o \simeq \hat{P}_-^o$, which is a good approximation since complete reflection occurs in the examples and the loss is low, then

$$\delta = \hat{P}_+^o(R_s - 1). \tag{12.38}$$

Curves of δ for the frequency of 3.16 mHz and the magnetic field direction of 30° and 60° are shown in Figure 12.8. The angle of incidence of the electromagnetic field is assumed to be the independent variable.

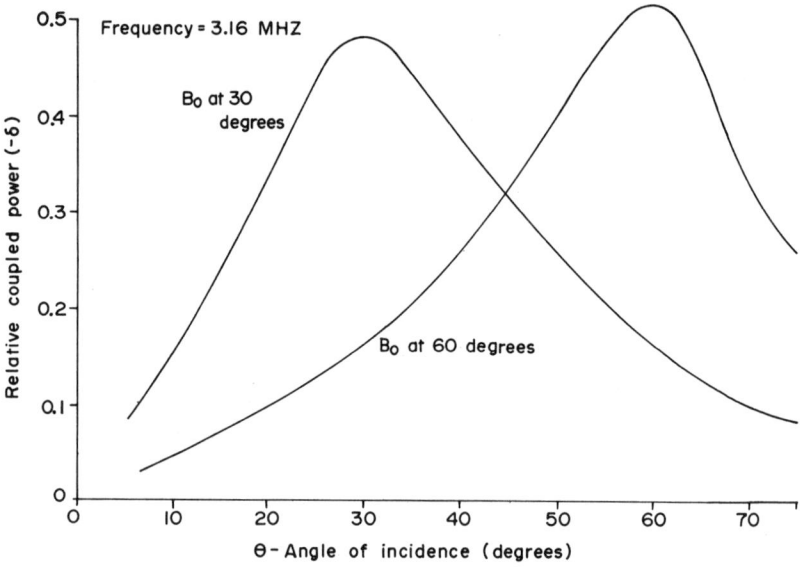

Figure 12.8. Relative coupled power vs. angle of incidence.

Figure 12.9 shows the relative coupled power function δ for 1.0 mHz and an incident angle of 60°. The coupled power reaches a maximum at 47°. Figure 12.10 is a plot of the ratio of the net coupled power to power in the extraordinary wave traveling in the direction of positive z vs. the angle of incidence. This curve seems to indicate that all of the power in the extraordinary wave is transferred to the ordinary wave when $\theta = 60°$.

The reader can draw his own conclusions on the results. It is difficult to present the data in such a way as to show clearly the coupling between the waves. None of the data obtained for the 10 mHz study have been presented. Interesting results were obtained, but time did not permit the study to be completed.

12 POLARIZATION OF AN ELECTROMAGNETIC WAVE

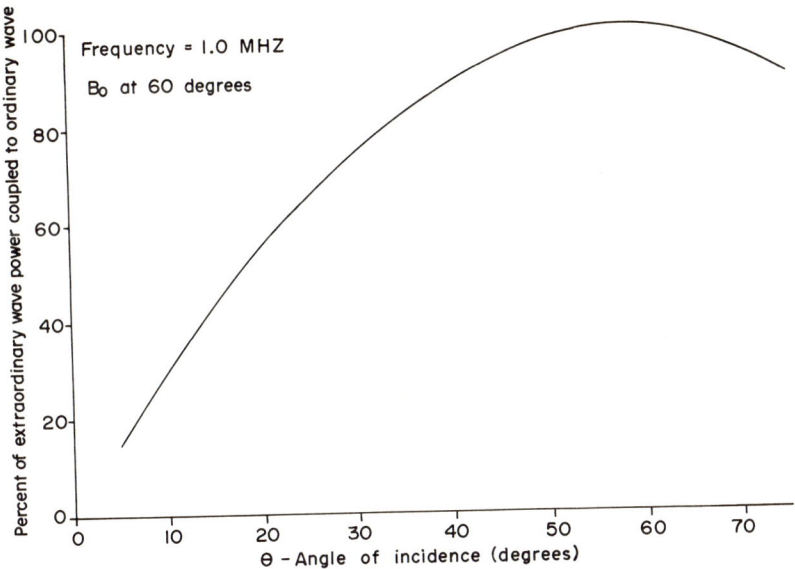

Figure 12.9. Relative coupled power vs. angle of incidence.

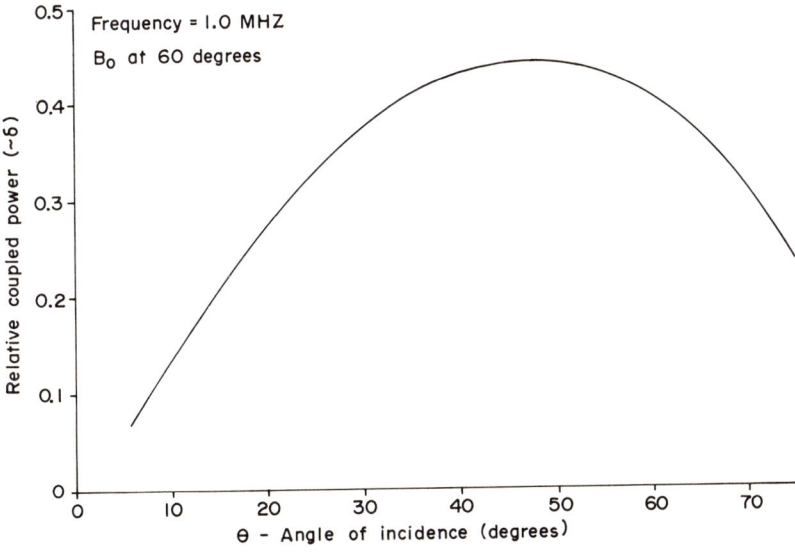

Figure 12.10. Percent of extraordinary wave power coupled to ordinary wave vs. angle of incidence.

6. SUMMARY

One method of studying the coupling between an ordinary and extraordinary wave in an inhomogeneous medium has been presented. The problem is a linear, two-point, boundary-value problem which can be attacked using the recursive scheme. The fundamental matrix was used to solve the differential equations; values of the fundamental matrix for the full medium were then used to find the initial conditions. Numerical checks were included in the program to provide some error analysis throughout in order to assure that the differential equations were being solved correctly.

The maximum coupling between the ordinary and extraordinary waves occurred when the angle of incidence of the wave from free space was approximately equal to the angle of magnetic field. The exact reason for this is not clear.

REFERENCES

1. H. A. Davies, Invariant Imbedding Techniques Used to Solve Wave Propagation Problems in a Horizontally Stratified Plasma Media, Ph.D. Dissertation, Vanderbilt University, 1969. Available from University Microfilms, Ann Arbor, Michigan.
2. J. A. Ratcliffe, *Magneto-Ionic Theory and its Application to the Ionosphere*, Cambridge University Press, London, 1959.
3. K. G. Budden, *Radio Waves in the Ionosphere*, Cambridge University Press, London, 1961.

CHAPTER 13

Electromagnetic-Electroacoustic Wave Coupling in a Compressible Plasma

The study of electromagnetic waves in an ionized medium is usually carried out under the assumption that the medium is an incompressible electron fluid. The motion of the ions is not considered in such an analysis. Such an assumption was made in Chapter 12.

To describe completely the behavior of an electromagnetic wave in an ionized medium, the motion of all charged particles which are free to move should be included. Maxwell's equations alone can never fully describe the behavior of the wave in an ionized medium. Since the charged particles are exposed to the electromagnetic field and experience a force, magnetohydrodynamic waves may be excited. Some of the magnetohydrodynamic waves which can be generated under certain conditions are the Alfvén, modified Alfvén, and acoustic waves. Coupling between these waves and the electromagnetic wave may exist such that energy can be exchanged (see Poeverlein [1] or Gerjuoy [2]).

This chapter discusses the coupling† between the electromagnetic wave and one of the acoustic waves, the electron pressure wave. The ionized medium is considered as a one-dimensional inhomogeneous medium with all ions in a "frozen" state. The ionized medium is assumed to be in a magnetically free region. This assumption does not have to be made; it is made to keep the number of modes to a minimum. The number of modes would also increase if the motion of the ions were included, one additional mode would have to be included for each species of ion present.

It may appear that some unnecessary restrictions have been imposed on the problem. This is partially true because the model is unrealistic. The purpose of simplifying the problem was to make the mathematics understandable and to illustrate the method.

The recursive equations to be used are those of the discrete version of the

---

† The material presented here has been extracted from an M.S. Thesis written by Windsor Letton, III; see [3].

numerical technique. The medium is cut into slabs of thickness Δz, each slab being homogeneous. There is a discontinuity in the medium characteristics at each interface. The transmission matrix at each interface must be found for the recursive calculations. A phase delay is included for each slab of thickness Δz.

The extension of the calculations to the continuously varying medium is straightforward. Chapter 12 described a continuous-type problem which the reader can use as a guide for solving the differential equations for the two waves.

I. THE IONIZED MEDIUM AND THE LINEARIZED PLASMA EQUATIONS

As mentioned in the introduction above, Maxwell's equations are necessary but not sufficient to describe fully the behavior of an electromagnetic wave in an ionized medium. The hydrodynamic equations of motion must also be imposed on the medium, as well as the conservation of charge. The equation of motion for an electron in a fluid is given by

$$\frac{\partial \bar{U}_e}{\partial t} + (\nabla \cdot \bar{U}_e)\bar{U}_e + \frac{e}{m_e}(\bar{E} + \bar{U}_e \times \bar{B}) + \frac{1}{N_e m_e}\nabla P_e + \frac{N_i}{m_e}\alpha \bar{U}_e = 0, \quad (13.1)$$

where

$\bar{U}_e$ = velocity of electrons,

$-e$ = electron charge,

m_e = rest mass of electron,

$N_e, (N_i)$ = number of electrons (ions) per unit volume,

P_e = partial pressure of electrons,

α = constant, dependent on scattering cross section,

$\bar{E}$ = electric field,

$\bar{H}$ = magnetic field

The continuity equation for the electrons is

$$\nabla \cdot (N_e \bar{U}_e) + \frac{\partial N_e}{\partial t} = 0, \quad (13.2)$$

where it is assumed that the ionized layer is neutralized by the electrons and ions.

13 ELECTROMAGNETIC-ELECTROACOUSTIC WAVE COUPLING

Equations (13.1) and (13.2) are not complete as they stand since the pressure of the electrons appears in (13.1). One additional equation, the partial pressure equation, must be defined. The partial pressure for the electrons is

$$P_e = N_e k T_e(N_e, N_i), \tag{13.3}$$

where k is the Boltzmann constant and T_e is the temperature of the electrons.

A set of equations for the ions, similar to those above, would be included if ions are to be considered. Since ions are so massive compared to electrons, they are neglected by making $\bar{U}_i = 0$.

Maxwell's equations must be modified when electrons and ions are present. The modified equations are

$$\nabla \times \bar{H} = -e\bar{U}_e + \varepsilon \frac{\partial \bar{E}}{\partial t}, \tag{13.4a}$$

$$\nabla \times \bar{E} = -\mu \frac{\partial \bar{H}}{\partial t}, \tag{13.4b}$$

$$\nabla \cdot \bar{E} = -\frac{e}{\varepsilon}(N_e - N_i), \tag{13.4c}$$

where μ is the magnetic permeability and ε is the dielectric constant for the medium.

The five equations given describe the electromagnetic wave behavior and the motion of the electrons in the ionized medium. The equations are not in the desired form for applying the recursive equation for a discrete medium. The exact form of the electromagnetic field which is incident on the ionized medium must be known to eliminate the operator $\nabla \times$ from (13.4a) and (13.4b). In addition, the equations must be linearized for the discrete medium. The linearization is carried out first. Assume that the electron velocity, the electron density, and the magnetic fields are composed of the sum of a constant average term and a time-varying perturbation term.

$$\bar{U}_e = \bar{U}_0 + \bar{u}_e, \tag{13.5a}$$

$$N_e = N_0 + n_e, \tag{13.5b}$$

$$\bar{H} = \bar{H}_0 + \bar{h}, \quad \bar{H}_0 = 0. \tag{13.5c}$$

The time-varying component is assumed to be small in comparison to the constant term. If isothermal conditions hold and if the mean collision frequency ν_e is equal to $N_e \alpha / m_e$, the linearized equations can be found by modifying the equations (13.1) to (13.4).

$$\rho_0 \left[\frac{\partial}{\partial t} + v_e \right] \bar{u}_e + \nabla p_e + \frac{e\rho_0}{m_e} \bar{E} = 0, \qquad (13.6a)$$

$$\rho_0 (\nabla \cdot \bar{u}_e) + \frac{\partial \rho}{\partial t} = 0, \qquad (13.6b)$$

$$\nabla p_e = v^2 \nabla \rho, \qquad (13.6c)$$

$$\nabla \times \bar{E} = -\mu \frac{\partial \bar{h}}{\partial t}, \qquad (13.6d)$$

$$\nabla \cdot \bar{H} = -\frac{\rho_0 \varepsilon}{m_e} \bar{u}_e + \varepsilon \frac{\partial \bar{E}}{\partial t}, \qquad (13.6e)$$

$$\nabla \cdot \bar{E} = \frac{-e}{m_e \varepsilon} \rho. \qquad (13.6f)$$

The six equations above are the desired linearized equations; ρ_0 is the average mass density, ρ is the varying mass density, and v is the adiabatic sound velocity.

Three waves can propagate in the ionized medium when (13.6) is valid. Two of the waves are the transverse electromagnetic waves with wave numbers

$$k = \frac{1}{c} \left[\omega^2 - \frac{\omega_e^2}{1 + v_e^2/\omega^2} - j \frac{v_e \omega_e^2}{\omega[1 + (v_e^2/\omega^2)]} \right]^{1/2}, \qquad (13.7)$$

where ω is the angular frequency of the wave, v_e is the collision frequency, and ω_e is the plasma frequency:

$$\omega_e = \left[\frac{N_0 e^2}{m_e \varepsilon_0} \right]^{1/2}, \qquad (13.8)$$

where N_0 is the average electron density. The third wave which can propagate is a longitudinal wave that is an electron pressure wave. This wave is the electroacoustic wave whose wave number is

$$K = \frac{1}{v} [(\omega^2 - \omega_e^2) - j\omega v_e]^{1/2}. \qquad (13.9)$$

The coupling between these three waves is examined in this chapter.

2. THE METHOD OF ANALYSIS

The coupled differential equations for the ordinary-extraordinary waves were derived and solved for the problem in the previous chapter. A slightly different approach is taken in the analysis of the electromagnetic-electroacoustic problem. Although the coupled differential equations can be solved here, the differential equations given in (13.6) are used to find the transmission, reflection, and coupling coefficients at an interface. The composite

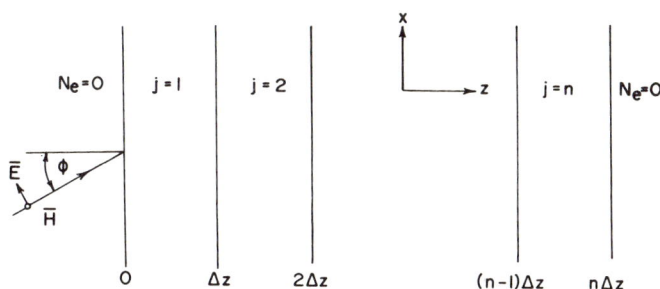

Figure 13.1. Stratified ionized medium.

effects of all of the interfaces, which have been created by slabbing the medium, are found by means of the recursive equations. If the slab thickness Δz is sufficiently small, the discrete recursive equations should give the correct composite transmission, reflection, and coupling coefficients.

To illustrate the method of analysis, consider the stratified medium shown in Figure 13.1. The electron distribution in the ionized medium is $N(z)$. The medium is broken into n slabs, each of thickness Δz and having an electron density $N_e(j)$, where $j = 1, 2, \ldots, n$. There will be $j + 1$ interfaces created where the electron density changes by a finite amount.

Let the electromagnetic wave be a plane wave which is incident on the ionized medium at an angle ϕ. The E field is restricted to the x-y plane with the H field perpendicular to the x-y plane. Only one component of E lies in the transverse plane, and only one transverse wave is considered. The medium is of infinite extent in the x-y plane.

It has been shown in an earlier paper [4] that an electroacoustic wave can be excited at a free space-plasma interface, as well as at a plasma-plasma interface, by an electromagnetic wave which has a component perpendicular to that interface. If $\phi \neq 0$, an electromagnetic wave component is present at the first interface at $z = 0$. A small amount of energy is coupled from the electromagnetic wave into the electroacoustic wave. It can also be shown that the reverse process occurs, an electroacoustic wave can produce an electromagnetic wave at the interface.

The determination of the transmission, reflection, and coupling coefficients at each interface is not a difficult task. There are boundary conditions which must hold at each interface, and these boundary conditions along with the equations of (13.6) provide the necessary means of finding the coefficients.

3. THE BOUNDARY CONDITIONS AND THE INTERFACE COEFFICIENTS

As mentioned earlier, three waves can propagate in the ionized medium. It is not difficult to show that there is an interrelationship between the velocity, electric field, magnetic field, and pressure. If superscript L denotes the longitudinal component and T the transverse component, it can be shown that

$$\bar{u}^L = j \frac{m\omega\varepsilon}{e\rho_0} \bar{E}^L, \tag{13.10a}$$

$$\bar{u}^T = j \frac{e}{m(\omega - jv)} \bar{E}^T, \tag{13.10b}$$

$$P = -\frac{mv^2\varepsilon}{e} \nabla \cdot \bar{E}^L, \tag{13.10c}$$

$$\bar{H}^T = \frac{j}{\omega\mu} \nabla \times E^T. \tag{13.10d}$$

These equations can be derived from the equations of (13.5) by considering the longitudinal and transverse components of the vector quantities.

The electric field, magnetic field, velocity, and pressure must satisfy certain boundary conditions at each interface throughout the medium. The electric and magnetic fields must have continuity of their tangential components across the interface. If the interface is in the x-y plane, then

$$\bar{a}_z \times (\bar{E}_i + \bar{E}_r - \bar{E}_t) = 0, \tag{13.11a}$$

$$\bar{a}_z \times (\bar{H}_i + \bar{H}_r - \bar{H}_t) = 0, \tag{13.11b}$$

where $\bar{a}_z$ is the unit vector along the z axis and the subscripts i, r, and t denote incident, reflected, and transmitted field components.

13 ELECTROMAGNETIC-ELECTROACOUSTIC WAVE COUPLING

The third boundary condition which can be imposed is that the net force on a boundary be zero:

$$P_i + P_r - P_t = 0. \tag{13.11c}$$

This implies that the interface is stationary.

The last boundary condition which must hold at an interface is that the normal component of the velocity be continuous:

$$\bar{a}_z \cdot (\bar{u}_i + \bar{u}_r - \bar{u}_t) = 0. \tag{13.11d}$$

The four relationships given in (13.10) along with the boundary conditions

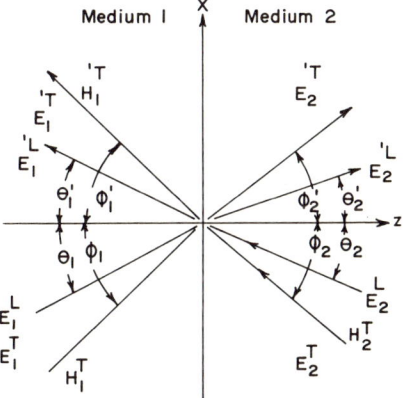

Figure 13.2. Wave branching at an interface in ionized medium.

of (13.11) are the equations required for finding the interface coefficients. The four equations of (13.10) are substituted into (13.11); the resulting equations are then used with the electric field components to find the coefficients.

Since there are two waves traveling in each direction within the medium, the transmission matrix is a 4 × 4 matrix with sixteen coefficients. To find these sixteen coefficients for a general interface, consider the interface shown in Figure 13.2. An electromagnetic wave, at an angle ϕ, is incident on each side of the interface. The electroacoustic wave, with an angle θ, is also incident on each side. The eight electric field components for the interface can be written down directly.

$$\bar{E}_{i1}^T = E_{i1}^T(-\bar{a}_z \sin \phi_{i1} + \bar{a}_x \cos \phi_{i1})\exp[j(\omega t - k_1 z \cos \phi_{i1} - k_1 x \sin \phi_{i1})], \tag{13.12a}$$

$$\bar{E}_{i1}^L = E_{i1}^L(\bar{a}_z \cos \theta_{i1} + \bar{a}_x \sin \theta_{i1})\exp[j(\omega t - K_1 z \cos \theta_{i1} - K_1 x \sin \theta_{i1})], \tag{13.12b}$$

$$\bar{E}_{i2}^T = E_{i2}^T(-\bar{a}_z \sin \phi_{i2} - \bar{a}_x \cos \phi_{i2})\exp[j(\omega t + k_2 z \cos \phi_{i2} - k_2 x \sin \phi_{i2})], \tag{13.12c}$$

$$\bar{E}_{i2}^L = E_{i2}^L(-\bar{a}_z \cos \theta_{i2} + \bar{a}_x \sin \theta_{i2})\exp[j(\omega t + K_2 z \cos \theta_{i2} - K_2 x \sin \theta_{i2})], \tag{13.12d}$$

$$\bar{E}_{r1}^T = E_{r1}^T(-\bar{a}_z \sin \phi_{r1} - \bar{a}_x \cos \phi_{r1})\exp[j(\omega t + k_1 z \cos \phi_{r1} - k_1 x \sin \phi_{r1})], \tag{13.12e}$$

$$\bar{E}_{r1}^L = E_{r1}^L(-\bar{a}_z \cos \theta_{r1} + \bar{a}_x \sin \theta_{r1})\exp[j(\omega t + K_1 z \cos \theta_{r1} - K_1 x \sin \theta_{r1})], \tag{13.12f}$$

$$\bar{E}_{r2}^T = E_{r2}^T(-\bar{a}_z \sin \phi_{r2} + \bar{a}_x \cos \phi_{r2})\exp[j(\omega t - k_2 z \cos \phi_{r2} - k_2 x \sin \phi_{r2})], \tag{13.12g}$$

$$\bar{E}_{r2}^L = E_{r2}^L(\bar{a}_z \cos \theta_{r2} + \bar{a}_x \sin \theta_{r2})\exp[j(\omega t - K_2 z \cos \theta_{r2} - K_2 x \sin \theta_{r2})]. \tag{13.12h}$$

Before deriving the sixteen coefficients of the transmission matrix, a few remarks should be made concerning the angular dependency of (13.12). It should be apparent that the incident and reflected angles for a particular type of wave at an interface are equal, i.e., $\theta_{i1} = \theta_{r1}$. It can also be shown that, regardless of how many times a wave undergoes reflections within the medium, the incident angle for that wave is always the same at a specified interface. This is true for the plane-parallel medium which is considered here.

Returning to the derivation of the coefficients, consider the incident wave $\bar{E}_{i1}^T$ alone with $\bar{E}_{i1}^L = \bar{E}_{i2}^T = \bar{E}_{i2}^L = 0$. The boundary condition

$$\bar{a}_z \times (\bar{E}_i + \bar{E}_r - \bar{E}_t) = 0 \tag{13.11a}$$

is considered first. Equations (13.12a) and (13.12e) to (13.12h) are still valid since the incident wave $\bar{E}_{i1}^T$ leads to reflected and transmitted transverse waves, as well as reflected and transmitted longitudinal waves at the interface. Substituting these expressions into (13.11a), dropping the time dependency, and setting z equal to zero, we find

$$1 - \frac{E_{r1}^T}{E_{i1}^T} + \frac{\sin \theta_{i1}}{\cos \phi_{i1}} \frac{E_{r1}^L}{E_{i1}^T} - \frac{\cos \phi_{i2}}{\cos \phi_{i1}} \frac{E_{r2}^T}{E_{i1}^T} - \frac{\sin \theta_{i2}}{\cos \phi_{i1}} \frac{E_{r2}^L}{E_{i1}^T} = 0. \quad (13.13a)$$

A second equation in the four unknowns of (13.13a) can be found by applying the boundary condition of (13.11b). Two additional equations can be derived by imposing the boundary conditions of (13.11c) and (13.11d). These equations are

$$1 - \frac{E_{r1}^T}{E_{i1}^T} - \frac{k_2}{k_1} \frac{E_{r2}^T}{E_{i1}^T} = 0, \quad (13.13b)$$

$$1 - \frac{E_{r1}^T}{E_{i1}^T} - \frac{K_1 \cos^2 \theta_{i1}}{k_1 \sin \phi_{i1} \cos \phi_{i1}} \frac{E_{r1}^L}{E_{i1}^T} - \frac{v_2^2 k_2 \sin \phi_{i2} \cos \phi_{i2}}{v_1^2 k_1 \sin \phi_{i1} \cos \phi_{i1}} \frac{E_{r2}^T}{E_{i1}^T}$$
$$+ \frac{v_2^2 K_2 \cos^2 \theta_{i2}}{v_1^2 k_1 \sin \phi_{i1} \cos \phi_{i1}} \frac{E_{r2}^L}{E_{i1}^T} = 0, \quad (13.13c)$$

$$1 + \frac{E_{r1}^T}{E_{i1}^T} + \frac{m^2 \omega \gamma_1 \varepsilon_0 \cos \theta_{i1}}{e^2 \rho_{01} \sin \phi_{i1}} \frac{E_{r1}^L}{E_{i1}^T} - \frac{\gamma_1 \sin \phi_{i2}}{\gamma_2 \sin \phi_{i1}} \frac{E_{r2}^T}{E_{i1}^T} + \frac{m^2 \omega \gamma_1 \varepsilon_0 \cos \theta_{i2}}{e^2 \rho_{02} \sin \phi_{i1}} \frac{E_{r2}^L}{E_{i1}^T} = 0,$$
$$(13.13d)$$

where $\gamma_m = \omega - i\nu_m$.

Four equations in the four unknowns are now available for finding four of the sixteen coefficients. The four unknowns of (13.13) can be identified from the transmission matrix. If $u(0)$ is the input vector for the left side of the interface and $v(x)$ the input vector for the right side, then the four coefficients obtained from (13.13) are the first row of the 4 × 4 matrix. The second row is found by making $\bar{E}_i^L$ finite and $\bar{E}_i^T = \bar{E}_i^L = \bar{E}_i^T = 0$. This process is carried out for each incident wave to obtain the complete transmission matrix at an interface.

The longitudinal wave cannot be propagated in a free-space region; therefore, a free space-plasma and a plasma-free space interface must be considered on each side of the ionized medium. The coefficients for interfaces having free space on one side can be derived by a procedure similar to the one above. There are only eight coefficients for such an interface.

The transmission matrix must be calculated for each interface in the medium. The sixteen equations which can be derived for the interface are valid for the entire medium provided that the incident angles, the wave number, the collision frequency, and other physical constants which appear in the equation are known. These constants will be known for a specified ionized medium.

4. THE IONIZED LAYER MODEL

The method of obtaining the transmission, reflection, and coupling coefficients has been outlined above. A specific ionized layer, the Epstein profile, was selected for the numerical analysis. This profile was discussed in Chapter 12. The electron density was specified as an Epstein profile in Chapter 12, where

$$N(z) = \frac{4M \exp[m(z - z_0)]}{1 + \exp[m(z - z_0)]^2}. \tag{13.14}$$

The value of M was selected to give the desired maximum electron density at $z = z_0$. The spread constant m was chosen as $35.25/\lambda_0$.

The ionized layer was fixed at a thickness of 80 cm. The layer was broken into 200 slabs, each being 0.4 cm thick. This slab thickness is small compared to the wavelength of the highest frequency selected. Frequencies of 2.5×10^8, 9.6×10^8, 22×10^8, and 55×10^8 Hertz were used in the calculations. The collision frequency was selected to be directly proportional to the electron density. Such an assumption is not always valid; it is difficult to assume any functional relationship which holds for a large number of cases. The temperature of the electrons was fixed at 4000°K.

5. NUMERICAL RESULTS

The recursive equations for the coupling between the electromagnetic-electroacoustic waves were programmed using the approach sketched above. The program was first tested for power conservation across the ionized layer by neglecting collision effects, that is, by making $\nu_e = 0$. If there is no loss in the medium, then the relation

$$R^2 + T^2 = 1$$

should be satisfied. This expression was satisfied with the deviation from unity in the seventh place. This accuracy was considered to be sufficient for the simplified model.

Upon completing the program test for a lossless medium, collisions were included in the calculations. The frequencies given above were used for the electromagnetic wave. The incident angle of the wave on the medium was varied from 10° to a maximum angle which is less than the "critical angle." This angle is approximately 75° for the ionized layer selected. Electron density maxima of 1.49×10^8 electrons/cm^3 and 10^{10} electrons/cm^3 were used in the sample calculations.

Typical data for the ionized layer are given in Table 13.1. The electron

Table 13.1. Imaginary Components of Electromagnetic and Electrostatic Wave Numbers

Frequency of (9.6×10^8 Hz)

Slab	Electron Density (elec./cm^3)	Collision Freq. (sec^{-1})	Im(k), EM wave (m^{-1})	Im(K) EA wave (m^{-1})
1	5.28×10^3	2.64×10^5	-2.01×10^{-10}	-5.36×10^{-1}
20	4.92×10^4	2.46×10^6	-1.74×10^{-8}	-5.00×10^0
40	5.16×10^5	2.57×10^7	-1.92×10^{-6}	-5.24×10^1
60	5.32×10^6	2.66×10^8	-2.04×10^{-4}	-5.40×10^2
80	4.74×10^7	2.36×10^9	-1.40×10^{-2}	-4.73×10^3
100	1.49×10^8	7.45×10^9	-6.35×10^{-2}	-1.33×10^4

Table 13.2. Reflection Coefficients for Lossy Epstein Layer: Maximum Electron Density = 1.49172×10^8 electrons/cc

	Frequency ($\times 10^6$ Hz)	
Angle	250	960
10°	$-1.591 \times 10^{-2} e^{-j79.5°}$	$-1.437 \times 10^{-3} e^{j19.7°}$
20°	$-1.527 \times 10^{-2} e^{-j35.8°}$	$-1.650 \times 10^{-3} e^{j61.5°}$
30°	$-1.982 \times 10^{-2} e^{j17.3°}$	$2.071 \times 10^{-3} e^{-j45.4°}$
40°	$-3.480 \times 10^{-2} e^{j66.6°}$	$2.964 \times 10^{-3} e^{j59.5°}$
50°	$6.270 \times 10^{-2} e^{-j70.3°}$	$-5.392 \times 10^{-3} e^{j12.0°}$
55°	$8.333 \times 10^{-2} e^{-j49.2°}$	$-8.026 \times 10^{-3} e^{j5.9°}$
60°	$1.103 \times 10^{-1} e^{-j27.6°}$	$1.247 \times 10^{-2} e^{-j20.8°}$
65°	—	$1.983 \times 10^{-2} e^{j55.6°}$
70°	—	$-3.204 \times 10^{-2} e^{-j46.3°}$
75°	—	$-5.299 \times 10^{-2} e^{j33.4°}$

Angle	2200	5500
10°	$-2.152 \times 10^{-5} e^{j9.7°}$	$6.074 \times 10^{-7} e^{-j36.8°}$
20°	$2.460 \times 10^{-5} e^{-j85.9°}$	$-6.643 \times 10^{-7} e^{j24.0°}$
30°	$3.428 \times 10^{-5} e^{j50.2°}$	$-5.136 \times 10^{-7} e^{j58.9°}$
40°	$-6.729 \times 10^{-5} e^{j61.8°}$	$3.897 \times 10^{-7} e^{j82.1°}$
50°	$-1.818 \times 10^{-4} e^{-j40.7°}$	$8.094 \times 10^{-7} e^{72.6°}$
55°	$3.427 \times 10^{-4} e^{-j68.2°}$	$-1.677 \times 10^{-6} e^{-j85.7°}$
60°	$-7.270 \times 10^{-4} e^{-j83.8°}$	$-3.316 \times 10^{-6} e^{-j41.3°}$
65°	$-1.712 \times 10^{-3} e^{j87.8°}$	$-4.542 \times 10^{-6} e^{j18.1°}$
70°	$4.267 \times 10^{-3} e^{j83.5°}$	$7.924 \times 10^{-6} e^{j64.7°}$
75°	$-1.085 \times 10^{-2} e^{j82.7°}$	$-1.800 \times 10^{-4} e^{-j20.1°}$

densities for slabs on the interior of the layer are given for a few of the slabs. The 100th slab corresponds to the middle of the ionized medium. The imaginary part of the wave numbers for the two waves as well as the assumed collision frequency for the particular slab are given.

The imaginary part of the wave vector is indicative of the attenuation which the wave undergoes in the slab. For example, if the slab thickness is Δz centimeters, the electroacoustic wave is attenuated by the real part of $\exp(-K_i \Delta z)$ for the ith slab. In the case of the 100th slab, the real part of $\exp(-jK_i \Delta z)$ is $\exp(-\alpha z)$; $-\alpha$ is given in Table 13.1 as -1.33×10^4. The energy which is coupled from the electromagnetic wave into the electroacoustic wave is highly attenuated in the center of the ionized medium.

The numerical results of the study are given in Tables 13.2, 13.3, 13.4, and 13.5. The reflection and transmission coefficients for the ionized layer

Table 13.3. Transmission Coefficients for Lossy Epstein Layer: Maximum Density = 1.49172×10^8 electrons/cc

	Frequency ($\times 10^6$ Hz)	
Angle	250	960
10°	$-0.98041e^{-j55.7°}$	$-0.99233e^{-j7.6°}$
20°	$-0.98139e^{-j44.4°}$	$-0.99268e^{j34.1°}$
30°	$-0.98286e^{-j26.0°}$	$0.99326e^{-j77.9°}$
40°	$-0.98452e^{-j0.9°}$	$0.99405e^{j14.6°}$
50°	$-0.98564e^{j30.4°}$	$-0.99502e^{-j51.4°}$
55°	$-6.98549e^{j48.2°}$	$-0.99555e^{j12.6°}$
60°	$-0.98427e^{j67.3°}$	$-0.99608e^{j80.8°}$
65°	—	$0.99658e^{-j27.4°}$
70°	—	$0.99689e^{j47.5°}$
75°	—	$-0.99664e^{-j54.7°}$
Angle	2200	5500
10°	$0.99762e^{j79.1°}$	$-0.99956e^{j16.8°}$
20°	$-0.99773e^{-j5.6°}$	$0.99958e^{j75.2°}$
30°	$0.99791e^{-j29.8°}$	$-0.99961e^{-j75.6°}$
40°	$-0.99815e^{j1.5°}$	$0.99965e^{-j64.5°}$
50°	$0.99845e^{j82.1°}$	$-0.99971e^{j24.0°}$
55°	$-0.99862e^{j48.5°}$	$-0.99974e^{j29.7°}$
60°	$0.99879e^{j24.2°}$	$-0.99977e^{j58.5°}$
65°	$-0.99898e^{j7.9°}$	$0.99981e^{-j72.5°}$
70°	$0.99918e^{-j1.5°}$	$0.99984e^{-j6.6°}$
75°	$-0.99938e^{-j5.3°}$	$0.99988e^{j73.2°}$

13 ELECTROMAGNETIC-ELECTROACOUSTIC WAVE COUPLING

Table 13.4. Reflection Coefficients for Lossy Epstein Layer: Maximum Electron Density = 10^{10} electrons/cc

Angle	Frequency ($\times 10^6$ Hz)	
	2200	5500
10°	$-2.218 \times 10^{-4} e^{-j18.9°}$	$-2.683 \times 10^{-7} e^{-j84.1°}$
20°	$-4.499 \times 10^{-4} e^{j74.2°}$	$-2.539 \times 10^{-7} e^{j46.2°}$
30°	$1.406 \times 10^{-3} e^{j72.8°}$	$-7.426 \times 10^{-7} e^{j16.8°}$
40°	$7.565 \times 10^{-3} e^{-j58.9°}$	$-1.196 \times 10^{-6} e^{-j19.7°}$
50°	$4.134 \times 10^{-2} e^{j31.2°}$	$-4.215 \times 10^{-6} e^{j66.9°}$
55°	$9.622 \times 10^{-2} e^{j6.5°}$	$1.440 \times 10^{-5} e^{j2.9°}$
60°	$-2.252 \times 10^{-1} e^{-j3.5°}$	$1.411 \times 10^{-4} e^{-j1.7°}$
65°	$4.916 \times 10^{-1} e^{j2.3°}$	$8.676 \times 10^{-4} e^{j51.6°}$
70°	—	$-7.456 \times 10^{-3} e^{-j54.7°}$
75°	—	$-6.199 \times 10^{-2} e^{j36.4°}$

Table 13.5. Transmission Coefficients for Lossy Epstein Layer: Maximum Electron Density = 10^{10} electrons/cc

Angle	Frequency ($\times 10^6$ Hz)	
	2200	5500
10°	$-0.9649 e^{-j89.2°}$	$-0.9716 e^{j43.7°}$
20°	$-0.9667 e^{j0.9°}$	$-0.9730 e^{-j78.0°}$
30°	$0.9695 e^{-j10.8°}$	$-0.9751 e^{-j49.8°}$
40°	$-0.9733 e^{j28.0°}$	$0.9780 e^{-j60.2°}$
50°	$-0.9774 e^{-j59.1°}$	$-0.9816 e^{j52.2°}$
55°	$0.9764 e^{-j83.2°}$	$-0.9836 e^{j59.2°}$
60°	$0.9580 e^{j86.5°}$	$0.9858 e^{-j0.05°}$
65°	$0.8571 e^{-j87.8°}$	$0.9881 e^{-j37.6°}$
70°	—	$0.9905 e^{j34.5°}$
75°	—	$-0.9912 e^{-j53.5°}$

with an Epstein profile are given in complex form. Four frequencies were allowed for the lower electron density, whereas only two frequencies are used for the higher electron density. Data are given for incident angles below the critical value only.

Curves of the transmissivity, defined as

$$\text{Transmissivity} = TT^*,$$

are given in Figures 13.3 and 13.4. Data for these curves were obtained from

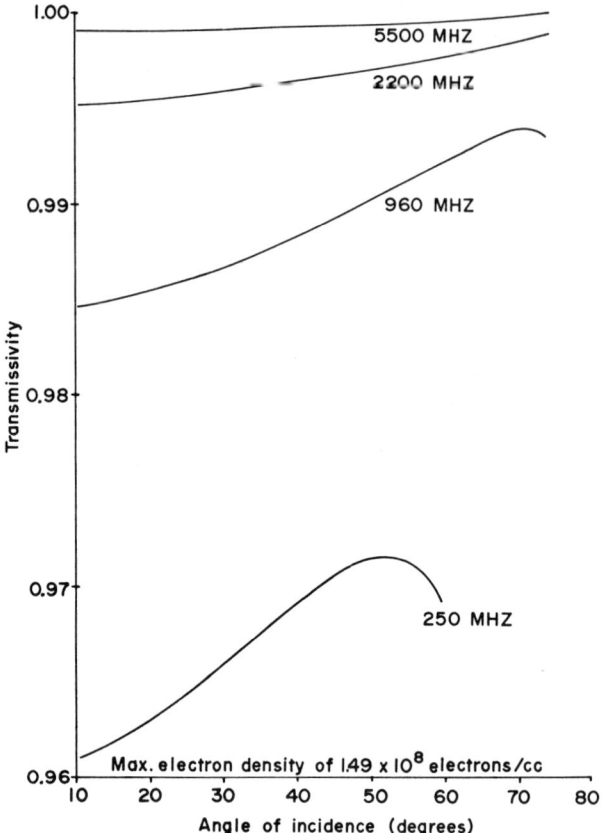

Figure 13.3. Transmissivity of electromagnetic wave vs. angle of incidence.

the tables. No attempt was made to obtain data near the critical angle; the angles used were given in Table 13.5.

The lossy medium has an absorption factor which can be defined as

$$\text{Absorption factor} = 1 - TT^* - RR^*.$$

Curves for the absorption factor, which is defined as the loss of energy, are given in Figure 13.5. There is a very small loss of energy for the lower electron density when the electromagnetic wave is of high frequency. The higher electron density absorbs a significant amount when the frequency is low. It should be kept in mind that the ionized layer is quite thin, only 200 cm in thickness.

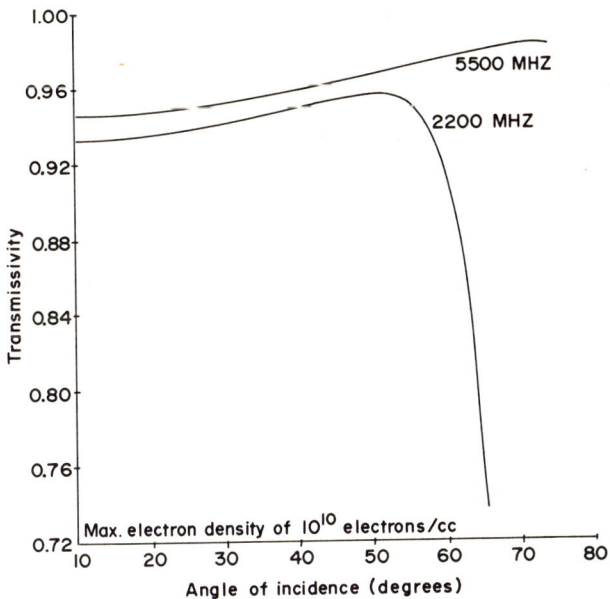

Figure 13.4. Transmissivity of electromagnetic wave vs. angle of incidence.

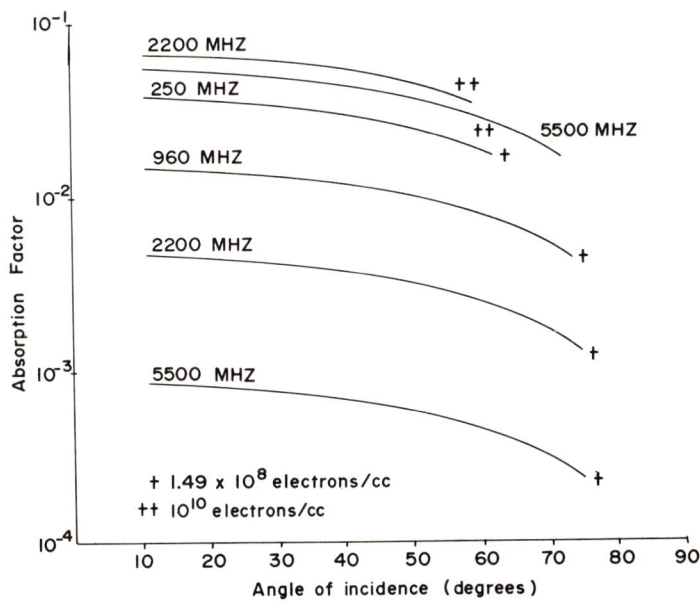

Figure 13.5. Absorption factor for lossy Epstein profile.

6. SUMMARY

The writer will let the reader draw his own conclusion about the data presented. The writer has tried to show that the recursive method is applicable to the electromagnetic-electroacoustic coupling problem. Many of the assumptions about the electron density and collision frequency dependency may be unrealistic. Selection of realistic models does not invalidate the numerical method.

The discrete recursive equations have been used; however, this is not necessary. The reason for using them here was that the discrete equations had been tested, whereas the continuous modeling was still being tested when the data above were obtained. The continuous recursive equations are perhaps more accurate and the "artificial" resonance effects, described in a paper by Adams and Denman, can be avoided.

Much work remains to be done to understand better the coupling problems in magnetohydrodynamics. A numerical method for the simplest coupling problem of this type has been presented. The extension to a larger number of modes is not difficult.

REFERENCES

1. H. Poeverlein, Coupling of Magnetohydrodynamic Waves in Stratified Media, *Phys. Rev.*, *136* (1964), A1605.
2. E. Gerjouy, Multiple-Wave Propagation and Causality, *Ann. Phys.*, *1* (1965), 1.
3. W. Letton, III., A Study of Wave Propagation through a Plasma Using Coupled-Mode Techniques, M.S. Thesis, Vanderbilt University, 1967. Available from University Microfilms, Ann Arbor, Michigan.
4. R. L. Gallawa, Propagation of Waves Across a Magnetoplasma-Vacuum Boundary, *Radio Sci.*, *69D* (1965), 807.

RELATED REFERENCES

A. H. Kritz and D. Mintzer, Propagation of Plasma Waves across a Density Discontinuity, *Phys. Rev.*, *117* (1960), 382.
L. Oster, Linearized Theory of Plasma Oscillations, *Rev. Mod. Phys*, *32* (1960), 141.

CHAPTER 14

Parametric Amplifiers

The parametric amplifier is a coupled mode device in which three or more electrical waves interact and exchange energy. The interaction usually takes place in a nonlinear element or because of a nonlinear characteristic of the medium. The interaction within a medium has been illustrated in the past few chapters. The coupled mode theory is applied to a very simple discrete type of parametric amplifier. Such a device as described would not be the usual type of parametric amplifier which would be designed for amplification purposes. It is presented here for illustrating the analysis. The extension to a useful parametric amplifier is left to the reader.

The parametric amplifier requires that a wave of large amplitude be present in the device to excite the nonlinear effect. This source is usually called the pump source. Let the pump source have a frequency ω_3. The signal which is amplified is generally of a low amplitude and at a frequency ω_1. The two waves, $E(\omega_1)$ and $E(\omega_3)$, interact in the nonlinear element to produce an infinite number of waves of frequencies $n\omega_3 \pm m\omega_1$, where n and m are integers. Many of these waves cannot propagate within the device and are ignored. A few of the secondary waves must be present to produce the energy exchange mechanism; thus not all of the waves can be ignored.

It is assumed that the pump source is connected directly to the nonlinear element and that waves with a frequency ω_3 or higher cannot propagate. The only waves that can propagate under this assumption are the signal wave $E(\omega_1)$ and a wave of frequency $\omega_3 - \omega_1$. The analysis is then restricted to a two-mode problem, the simplest that can be considered. The wave at a frequency $\omega_2 = \omega_3 - \omega_1$ is called the idler wave.

In addition to the restrictions placed on the number of modes above, some further restrictions are placed on the device. The phase relationship between the pump, signal, and idler waves is neglected. This phase relationship is quite important in making a device work. The only phase that is introduced into the amplifier is that due to the transmission lines which connect the various elements together.

The writer realizes that the model has been simplified to an almost trivial case; the reason for some of this simplification will be apparent when the gain equations are inspected. The purpose of the discussion is to show how the method is applied, not to consider all of the variables that can enter into the equations. The gain equations are sufficiently complicated with these assumptions without including all of the other effects.

I. SIMPLE AMPLIFIER

The discrete amplifier to be discussed in this chapter is the one shown in Figure 14.1. The parametric device consists of five sections, which are

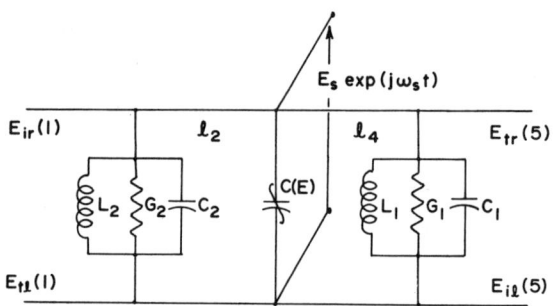

Figure 14.1. Simple parametric amplifier.

numbered from left to right. The device is a two-port network consisting of a voltage-sensitive nonlinear capacitor and separated from two resonant circuits by two transmission lines of length l_2 and l_4. The resonant circuit on the left side, section (1), is tuned to the idler frequency ω_2.

The resonant circuit on the right side, section (5), is tuned to the frequency ω_1, where $\omega_1 < \omega_2 < \omega_3$. The two resonant circuits are assumed to have very large quality factors, Q. The quality factor is not to be confused with the fundamental matrix $\mathbf{Q}(0, x)$.

The analysis of the parametric device consists in finding the transmission matrix for each section and then using the recursive equations for cascading. The composite transmission matrix describes the performance of the device.

The equations for the wave for each section, j, can be written as in (14.1). The signal flow graph

$$\begin{bmatrix} E_{tr}(j) \\ E_{tl}(j) \end{bmatrix} = \begin{bmatrix} \tau(j) & R(j) \\ \mathcal{R}(j) & T(j) \end{bmatrix} \begin{bmatrix} E_{ir}(j) \\ E_{il}(j) \end{bmatrix} \tag{14.1}$$

14 PARAMETRIC AMPLIFIERS

for each section can be drawn from (14.1) if desired. The reader should keep in mind that the field parameters are column vectors of order 2. For example,

$$E_{ir}(j) = \begin{bmatrix} E_{ir}^s(j) \\ E_{ir}^i(j) \end{bmatrix}, \tag{14.2}$$

where the superscripts denote the signal frequency ω_1 and the idler frequency ω_2. The remaining three electric fields of (14.1) can be written in a similar way. The argument j takes on the values 1, 2, ..., 5 for the five sections.

The transmission matrix for each section is a 4 × 4 matrix with scalar coefficients. The entries

$$\mathbf{T}(j) = \begin{bmatrix} \tau_{11}(j) & \tau_{12}(j) & R_{11}(j) & R_{12}(j) \\ \tau_{21}(j) & \tau_{22}(j) & R_{21}(j) & R_{22}(j) \\ \mathscr{R}_{11}(j) & \mathscr{R}_{12}(j) & T_{11}(j) & T_{12}(j) \\ \mathscr{R}_{21}(j) & \mathscr{R}_{22}(j) & T_{21}(j) & T_{22}(j) \end{bmatrix} \tag{14.3}$$

of the transmission matrix may be complex numbers. If there is no interaction between modes, the off-diagonal elements of $\tau(j)$, $R(j)$, $\mathscr{R}(j)$, and $T(j)$ are zeros. The only interaction permitted in this model takes place in the non-linear capacitor.

It has been shown earlier that the transmission and scattering matrices are related. Equation (14.4),

$$\begin{bmatrix} \tau(j) & R(j) \\ \mathscr{R}(j) & T(j) \end{bmatrix} = \begin{bmatrix} 0 & I \\ I & 0 \end{bmatrix} \begin{bmatrix} S_{11}(j) & S_{12}(j) \\ S_{21}(j) & S_{22}(j) \end{bmatrix}, \tag{14.4}$$

applies provided that the waves are written as vectors. Consider the two-port network shown in Figure 2.3. The two-port network is considered without interaction and the scattering matrix can be written down in partitioned form.

$$\begin{bmatrix} E_{il}^s(j) \\ E_{tr}^s(j) \\ E_{il}^i(j) \\ E_{tr}^i(j) \end{bmatrix} = \begin{bmatrix} S_{11}^s & S_{12}^s & 0 & 0 \\ S_{21}^s & S_{22}^s & 0 & 0 \\ 0 & 0 & S_{11}^i & S_{12}^i \\ 0 & 0 & S_{21}^i & S_{22}^i \end{bmatrix} \begin{bmatrix} E_{ir}^s(j) \\ E_{il}^s(j) \\ E_{ir}^i(j) \\ E_{il}^i(j) \end{bmatrix} \tag{14.5}$$

Because, from (14.2), the desired vectors are not present, equation (14.5) must be rearranged. Rearranging (14.5) and applying the transformation given in (14.4), the transmission matrix is obtained. The transmission matrix

$$\mathbf{T}(j) = \begin{bmatrix} S^s_{21} & 0 & S^s_{22} & 0 \\ 0 & S^i_{21} & 0 & S^i_{22} \\ S^s_{11} & 0 & S^s_{12} & 0 \\ 0 & S^i_{11} & 0 & S^i_{12} \end{bmatrix} \qquad (14.6)$$

has the zeros; thus it indicates the absence of coupling between modes. If the elements of the scattering matrix are known, the transmission matrix can be written down directly. The transmission matrices for the five sections are now derived using the procedure above.

The two sections of transmission line can be analyzed without difficulties. The transmission line, when matched with its characteristic impedance, has transmission coefficients of unity magnitude. There is a phase shift $\beta_i l_j$, where l_j is the length of the line, for each section. There are no reflected waves, so all off-diagonal terms of (14.6) are zero. The transmission matrix for the line is

$$\mathbf{T}(j) = \begin{bmatrix} \exp(-j\beta_s l_j) & 0 & 0 & 0 \\ 0 & \exp(-j\beta_i l_j) & 0 & 0 \\ 0 & 0 & \exp(-j\beta_s l_j) & 0 \\ 0 & 0 & 0 & \exp(-j\beta_i l_j) \end{bmatrix}, \qquad (14.7)$$

where β_s and β_i are the wave numbers for the signal and idler waves and l_j is the length of line. The two transmission lines are sections (2) and (4); thus j takes on the values 2 and 4.

The transmission matrices for the two resonant circuits are not so easy to find as the transmission matrices for the line. The response of the circuit to two frequencies must be considered. The scattering matrix for the resonant circuit is not derived here; the matrix has been found by other workers and it can be written down directly. The scattering matrix for a two-port cavity for an excitation wave of frequency ω is

$$[S] = \frac{1}{1 + jX(\omega)} \begin{bmatrix} -jX(\omega) & \pm 1 \\ \pm 1 & -jX(\omega) \end{bmatrix}, \qquad (14.8)$$

where

$$X(\omega) = \frac{2\, d\omega}{\Delta\omega}. \qquad (14.9)$$

The function $X(\omega)$ depends on the difference $d\omega$ between the frequency of the

14 PARAMETRIC AMPLIFIERS

wave, ω, and the resonant frequency of the cavity, ω_0.

$$d\omega = \omega_0 - \omega. \tag{14.10}$$

The denominator of (14.9) is the 3db band width of the loaded cavity. This factor depends on the losses in the cavity or the quality factor. When the wave incident on the cavity is identical with the resonant frequency, $X(\omega_0) = 0$. If the incident wave is much higher or lower than the resonant frequency, $X(\omega)$ approaches $\pm \infty$. It is assumed that one of the modes is sufficiently off the resonant frequency that the value of infinity can be used. The exact value can be used when it is known.

The partitioned scattering matrix for the two modes is given in (14.9). The scattering matrix written in this way does not have the incident waves arranged in the column vector form. The sign of the off-diagonal terms,

$$[S] = \begin{bmatrix} \dfrac{-jX(\omega)}{1+jX(\omega)} & \dfrac{\pm 1}{1+jX(\omega)} & 0 & 0 \\ \dfrac{\pm 1}{1+jX(\omega)} & \dfrac{-jX(\omega)}{1+jX(\omega)} & 0 & 0 \\ 0 & 0 & \dfrac{-jX(\omega)}{1+jX(\omega)} & \dfrac{\pm 1}{1+jX(\omega)} \\ 0 & 0 & \dfrac{\pm 1}{1+jX(\omega)} & \dfrac{-jX(\omega)}{1+jX(\omega)} \end{bmatrix}, \tag{14.11}$$

depends on the electrical length of the cavity; the minus sign holds when the cavity is an even number of half-wavelengths long, and the positive sign holds when the cavity is an odd number of half-wavelengths.

Assume that the cavity in section (1) is tuned to the frequency ω_2 and that in cavity (5) to the frequency ω_1. Since $\omega_1 < \omega_2$, $X(\omega)$ for the frequency ω_1 is ∞ and $X(\omega)$ for ω_2 is zero. The transmission matrix for section (1) is then composed of zeros and minus ones.

$$T(1) = \begin{bmatrix} 0 & 0 & -1 & 0 \\ 0 & -1 & 0 & 0 \\ -1 & 0 & 0 & 0 \\ 0 & 0 & 0 & -1 \end{bmatrix}. \tag{14.12}$$

The minus sign has been selected which is for a cavity of even half-wavelengths.

The transmission matrix for section (5) can also be obtained from (14.11). If the circuit is tuned to ω_1, then $X(\omega)$ is $-\infty$ for the frequency ω_2 and $X(\omega)$ is zero for the frequency ω_1. The minus sign for $X(\omega)$ at the frequency arises becauses $\omega_2 \gg \omega_1$.

$$\mathbf{T}(5) = \begin{bmatrix} -1 & 0 & 0 & 0 \\ 0 & 0 & 0 & -1 \\ 0 & 0 & -1 & 0 \\ -1 & 0 & 0 & 0 \end{bmatrix}. \tag{14.13}$$

Because the nonlinear capacitor has been discussed earlier, the analysis is not repeated here. The circuit of Figure 14.1 has a shunt element; thus that form of transmission matrix should be used. The transmission matrix for the shunt element has been discussed in Chapter 9.

The five sections of the parametric device have been described by the individual transmission matrices. The gain of the device can be obtained by cascading the five sections together. It should be kept in mind that each section was analyzed with the assumption that no reflected waves occur, that is, the section is properly terminated. If there is an impedance change, this change must be taken into account. If the device is not terminated in a matched load, the simple formula for a step impedance change must be used.

2. NEGATIVE RESISTANCE, UP CONVERSION AND DOWN CONVERSION GAIN EQUATIONS

The circuit shown in Figure 14.1 can be operated in three different modes. This device can be used as a negative resistance amplifier, an up converter, or a down converter. The choice among these three modes depends on the application for the device.

The negative resistance amplifier is a device whose input and output frequencies are the same. Since the cavity at (1) is a short to a frequency ω_1, the input to the left-hand port must be a frequency ω_2. Similarly, the input to the right-hand port must be at a frequency ω_1. If ω_1 is to be the signal wave, then the right-hand port must serve as an input and an output port. The gain for the negative resistance amplifier is then given by the ratio of the output wave to the input wave at the right-hand port. The composite transmission matrix for the device is $\mathbf{T}(1, 2, 3, 4, 5)$. The matrix equation for the waves

14 PARAMETRIC AMPLIFIERS

can then be written as

$$\begin{bmatrix} E_{tr}^s \\ E_{tr}^i \\ E_{tl}^s \\ E_{tl}^i \end{bmatrix} = \mathbf{T}(1, 2, 3, 4, 5) \begin{bmatrix} E_{ir}^s \\ E_{ir}^i \\ E_{il}^s \\ E_{il}^i \end{bmatrix}. \tag{14.14}$$

The gain of the negative resistance amplifier when operated with a signal frequency ω_1 is then given by the ratio E_{tr}^s/E_{il}^s. The magnitude of the gain is then the ratio of the transmitted wave moving in the right-hand direction to the incident wave moving in the left-hand direction, both measured at the right-hand port. It is now a simple task to determine which coefficient of the composite transmission matrix gives the negative resistance gain. The negative resistance gain is given by

$$\text{Negative resistance gain} = G_{NR}(\omega_1) = \frac{E_{tr}^s}{E_{il}^s} = R_{11}(1, 2, 3, 4, 5). \tag{14.15}$$

Equation (14.15) is valid when all other inputs are zero.

The up converter is a device that accepts an input at a low frequency and converts it into a signal at a higher frequency. The device may or may not have gain when operated in this mode. If $\omega_1 < \omega_2$, then the up converter gain is the ratio of the output wave at a frequency ω_2 to the input wave at a frequency ω_1. The port on the left side of the device of Figure 14.1 completely reflects an input signal at a frequency ω_1; thus the port on the right side must be the input port. The resonant circuit on the right does not permit a wave at a frequency ω_2 to pass; therefore the output port for the wave at a frequency ω_2 must be the left port. Using this reasoning, the up converter gain can be found from (14.14). The up converter gain is the ratio E_{tl}^i/E_{il}^s or

$$\text{Up converter gain} = G_{UC}(\omega_2) = \frac{E_{tl}^i}{E_{il}^s} = T_{21}(1, 2, 3, 4, 5). \tag{14.16}$$

The parametric device may be operated in a down-conversion mode, the signal input may be at a frequency ω_2 and emerge at a frequency ω_1. The input port for the wave at a frequency ω_2 must be the right-hand port. The gain is then given by ratio E_{tr}^s/E_{ir}^i.

$$\text{Down converter gain} = G_{DC}(\omega_1) = \frac{E_{tr}^s}{E_{ir}^i} = \tau_{12}(1, 2, 3, 4, 5). \tag{14.17}$$

Equations (14.15), (14.16), and (14.17) give the gain for the three possible modes of operation. If the five sections of the parametric device are cascaded together, the cascading algebra makes available the three gain expressions. The composite transmission matrix, **T**(1, 2, 3, 4, 5), makes it possible to examine such a device for any of these modes. The composite matrix is given in the next section.

3. THE GAIN EQUATIONS

The gain equations for $G_{NR}(\omega_1)$, $G_{UC}(\omega_2)$, and $G_{DC}(\omega_1)$ are now derived with the nonlinear capacitor expressed through a general form of the transmission matrix. Assume that the nonlinear capacitor has the transmission matrix

$$\mathbf{T}(3) = \begin{bmatrix} \tau_{11}(3) & \tau_{12}(3) & R_{11}(3) & R_{12}(3) \\ \tau_{12}(3) & \tau_{22}(3) & R_{21}(3) & R_{22}(3) \\ \mathscr{R}_{11}(3) & \mathscr{R}_{12}(3) & T_{11}(3) & T_{12}(3) \\ \mathscr{R}_{21}(3) & \mathscr{R}_{22}(3) & T_{21}(3) & T_{22}(3) \end{bmatrix}. \quad (14.18)$$

The sixteen coefficients of (14.18) depend on the static capacitance of the nonlinear capacitor, the pump, signal and idler frequencies, and the nonlinear characteristics of the capacitor.

If the transmission line on the left side of the circuit of Figure 14.1 is infinite in extent, **T**(1) is the cascaded transmission matrix for the input line and the resonant circuit. The recursive equations of Chapter 3 are used to cascade **T**(1) and **T**(2). The composite matrix **T**(1, 2) for the resonant circuit of frequency ω_2 and the transmission line of length l_2 is

$$\mathbf{T}(1,2) = \begin{bmatrix} 0 & 0 & -\exp(-2j\beta_1 l_2) & 0 \\ 0 & -\exp(-j\beta_2 l_2) & 0 & 0 \\ -1 & 0 & 0 & 0 \\ 0 & 0 & 0 & -\exp(-j\beta_2 l_2) \end{bmatrix}, \quad (14.19)$$

where $\beta_1 = 2\pi/\lambda_1$, and $\beta_2 = 2\pi/\lambda_2$.

The transmission matrix for the nonlinear capacitor **T**(3) can now be cascaded with **T**(1, 2) to find the composite matrix **T**(1, 2, 3). This composite

14 PARAMETRIC AMPLIFIERS

matrix is the transmission matrix for the first three sections in cascade,

$$\mathbf{T}(1,2,3) = \begin{bmatrix} 0 & -\tau_{12}(3)\delta_2 + \dfrac{\tau_{11}(3)\mathscr{R}_{12}(3)\delta_2\delta_1^2}{1+\mathscr{R}_{11}(3)\delta_1^2} & R_{11}(3) - \dfrac{\tau_{11}(3)T_{11}(3)\delta_1^2}{1+\mathscr{R}_{11}(3)\delta_1^2} & R_{12}(3) - \dfrac{\tau_{11}(3)T_{12}(3)\delta_1^2}{1+\mathscr{R}_{11}(3)\delta_1^2} \\ 0 & -\tau_{22}(3)\delta_2 + \dfrac{\tau_{21}(3)\mathscr{R}_{12}(3)\delta_2\delta_1^2}{1+\mathscr{R}_{11}(3)\delta_1^2} & R_{21}(3) - \dfrac{\tau_{21}(3)T_{11}(3)\delta_1^2}{1+\mathscr{R}_{11}(3)\delta_1^2} & R_{22}(3) - \dfrac{\tau_{21}(3)T_{12}(3)\delta_1^2}{1+\mathscr{R}_{11}(3)\delta_1^2} \\ -1 & 0 & 0 & 0 \\ 0 & \mathscr{R}_{22}(3)\delta_2^2 - \dfrac{\mathscr{R}_{21}(3)\mathscr{R}_{12}(3)\delta_1^2\delta_2^2}{1+\mathscr{R}_{11}(3)\delta_1^2} & -T_{21}(3)\delta_2 + \dfrac{T_{11}(3)\mathscr{R}_{21}(3)\delta_2\delta_1^2}{1+\mathscr{R}_{11}(3)\delta_1^2} & -T_{22}(3)\delta_2 + \dfrac{T_{12}(3)\mathscr{R}_{21}(3)\delta_2\delta_1^2}{1+\mathscr{R}_{11}(3)\delta_1^2} \end{bmatrix},$$

(14.20)

where

$$\delta_1 = \exp(-j\beta_1 l_2), \quad (14.21\text{a})$$

$$\delta_2 = \exp(-j\beta_2 l_2). \quad (14.21\text{b})$$

Equation (14.20) is now simplified by writing the coefficients in terms of the composite coefficients.

$$\mathbf{T}(1,2,3) = \begin{bmatrix} 0 & \tau_{12}(1,2,3) & R_{11}(1,2,3) & R_{12}(1,2,3) \\ 0 & \tau_{22}(1,2,3) & R_{21}(1,2,3) & R_{22}(1,2,3) \\ -1 & 0 & 0 & 0 \\ 0 & \mathscr{R}_{22}(1,2,3) & T_{21}(1,2,3) & T_{22}(1,2,3) \end{bmatrix}. \quad (14.22)$$

The cascading of the last two sections is carried out with these generalized coefficients. The expressions given in (14.20) would make the algebra unduly involved.

Before cascading $\mathbf{T}(1, 2, 3)$ to $\mathbf{T}(4)$ and $\mathbf{T}(5)$ in that order, the composite matrix $\mathbf{T}(4, 5)$ is found. The cascading of composite matrices is valid, and in this case some reduction in the algebra is possible. The transmission matrix

T(4) is given by (14.7), where $j = 4$, and T(5) is given in (14.13). The composite matrix T(4, 5) is

$$\mathbf{T}(4, 5) = \begin{bmatrix} -\exp(-j\beta_1 l_4) & 0 & 0 & 0 \\ 0 & 0 & 0 & 0 \\ 0 & 0 & -\exp(-j\beta_1 l_4) & 0 \\ 0 & -\exp(-2j\beta_2 l_4) & 0 & 0 \end{bmatrix}. \quad (14.23)$$

This is the composite matrix for the five sections in cascading the last two sections with the first three sections. The composite transmission, T(1, 2, 3, 4, 5), is

$$\mathbf{T}(1, 2, 3, 4, 5) = \begin{bmatrix} 0 & \tau_{12}(1,2,3,4,5) & R_{11}(1,2,3,4,5) & 0 \\ 0 & 0 & 0 & -1 \\ -1 & 0 & 0 & 0 \\ 0 & \mathscr{R}_{22}(1,2,3,4,5) & T_{21}(1,2,3,4,5) & 0 \end{bmatrix}. \quad (14.24)$$

The four elements which are not constants are the desired gain functions.

$$\tau_{12}(1, 2, 3, 4, 5) = -\tau_{12}(1, 2, 3)\delta_3 + \frac{R_{12}(1, 2, 3)\tau_{22}(1, 2, 3)\delta_3 \delta_3^2}{1 + R_{22}(1, 2, 3)\delta_4^2}, \quad (14.25a)$$

$$\mathscr{R}_{22}(1, 2, 3, 4, 5) = \mathscr{R}_{22}(1, 2, 3)\delta_3^2 - \frac{R_{12}(1, 2, 3)R_{21}(1, 2, 3)\delta_3^2 \delta_4^2}{1 + R_{22}(1, 2, 3)\delta_4^2}, \quad (14.25b)$$

$$R_{11}(1, 2, 3, 4, 5) = R_{11}(1, 2, 3)\delta_3^2 - \frac{R_{12}(1, 2, 3)R_{21}(1, 2, 3)\delta_3^2 \delta_4^2}{1 + R_{22}(1, 2, 3)\delta_4^2}, \quad (14.25c)$$

$$T_{21}(1, 2, 3, 4, 5) = -T_{21}(1, 2, 3)\delta_3 + \frac{T_{22}(1, 2, 3)R_{21}(1, 2, 3)\delta_3^2 \delta_4^2}{1 + R_{22}(1, 2, 3)\delta_4^2}. \quad (14.25d)$$

The composite coefficients for the first three sections are given in the matrix of (14.20). The phase shifts in the transmission line of section (4) are given by δ_3 and δ_4, where

$$\delta_3 = \exp(-j\beta_1 l_4), \quad (14.26a)$$

$$\delta_4 = \exp(-j\beta_2 l_4). \quad (14.26b)$$

There are six nonzero entries in the transmission matrix for the device. The two entries which have the value minus one are the reflection coefficients for the waves incident on cavities which are tuned to other frequencies. The four other terms are the two gain equations for the amplifier in a negative resistance mode, one for ω_1 and one for ω_2, and the up and down converter

14 PARAMETRIC AMPLIFIERS

gains. The gain equation for the negative resistance mode is

$$G_{NR}(\omega_1) = \left[R_{11} - \frac{\tau_{11}T_{12}\exp(-2j\beta_1 l_2)}{1 + \mathscr{R}_{11}\exp(-2j\beta_1 l_2)} \right]\exp(-2j\beta_1 l_4),$$

$$- \frac{\left[R_{12} - \dfrac{\tau_{11}T_{12}\exp(-2j\beta_1 l_2)}{1 + \mathscr{R}_{11}\exp(-2j\beta_1 l_2)} \right]\left[R_{21} - \dfrac{\tau_{21}T_{11}\exp(-2j\beta_1 l_4)}{1 + \mathscr{R}_{11}\exp(-2j\beta_1 l_2)} \right]}{1 + \left[R_{22} - \dfrac{\tau_{21}T_{12}\exp(-2j\beta_1 l_2)}{1 + \mathscr{R}_{11}\exp(-2j\beta_1 l_2)} \right]\exp(-2j\beta_2 l_4)}$$

$$\times \exp(-2j\beta_1 l_4)\exp(-2j\beta_2 l_4). \quad (14.27)$$

where R_{11}, τ_{11}, etc are transmission matrix elements for the nonlinear capacitor.

The other gain relationships can be written down from (14.25) and (14.20) if desired. Since the equations are quite long, the expressions are not given here. The reader can find them if they are desired.

The equations above have been derived under some simplifying assumptions that are quite restrictive. There has been no consideration of the phase relationships which may exist between the pump source and the signal wave. The author does not know of any reasons why they could not be considered in the analysis. Other types of parametric devices could be analyzed by the same procedure.

4. SUMMARY

The method outlined for analyzing parametric devices has not received sufficient attention to justify drawing conclusions. A more complete examination of the method is needed. The equations were being investigated by Mr. James S. Davis at the time of his death. Some preliminary calculations had been made, and the numerical results indicated that gain would occur for certain ranges of the variables. The calculations were very limited and no general conclusions were reached at that time.

APPENDIX A

Fourth-Order Incremental Coefficients

The fourth-order incremental coefficients for the transmission matrix are given below. These coefficients have been derived from the differential equations for the transmission matrix. The incremental coefficients can be calculated from the equation

$$\mathbf{T}(x, x + \Delta x) = \mathbf{T}_0(x, x) + \mathbf{T}_1(x, x)\Delta x + \mathbf{T}_2(x, x)\frac{\Delta x^2}{2}$$
$$+ \mathbf{T}_3(x, x)\frac{\Delta x^3}{6} + \mathbf{T}_4(x, x)\frac{\Delta x^4}{24}, \quad \text{(A-1)}$$

where the subscript indicates the order of the derivative evaluated for $\Delta x = 0$.

$$\mathbf{T}_0(x, x) = \begin{bmatrix} I & 0 \\ 0 & I \end{bmatrix}, \quad \text{(A-2)}$$

$$\mathbf{T}_1(x, x) = \begin{bmatrix} B(x) & A(x) \\ D(x) & C(x) \end{bmatrix} = \begin{bmatrix} T_1(x, x) & R_1(x, x) \\ \mathscr{R}_1(x, x) & T_1(x, x) \end{bmatrix}, \quad \text{(A-3)}$$

$$\mathbf{T}_2(x, x) = \begin{bmatrix} B'(x) & A'(x) \\ D'(x) & C'(x) \end{bmatrix} + \begin{bmatrix} B(x) & A(x) \\ D(x) & C(x) \end{bmatrix}\begin{bmatrix} T_1(x, x) & R_1(x, x) \\ \mathscr{R}_1(x, x) & T_1(x, x) \end{bmatrix}, \quad \text{(A-4)}$$

$$\mathbf{T}_3(x, x) = \begin{bmatrix} B''(x) & A''(x) \\ D''(x) & C''(x) \end{bmatrix} + 2\begin{bmatrix} B'(x) & 0 \\ D'(x) & 0 \end{bmatrix}\begin{bmatrix} T_1(x, x) & R_1(x, x) \\ 0 & 0 \end{bmatrix}$$
$$+ 2\begin{bmatrix} 0 & R_1(x, x) \\ 0 & T_1(x, x) \end{bmatrix}\begin{bmatrix} 0 & 0 \\ D'(x) & C'(x) \end{bmatrix}$$
$$+ 2\begin{bmatrix} 0 & R_1(x, x) \\ 0 & T_1(x, x) \end{bmatrix}\begin{bmatrix} 0 & 0 \\ \mathscr{R}_1(x, x) & 0 \end{bmatrix}$$
$$\cdot \begin{bmatrix} T_1(x, x) & R_1(x, x) \\ 0 & 0 \end{bmatrix} + \begin{bmatrix} T_1(x, x) & 0 \\ \mathscr{R}_1(x, x) & 0 \end{bmatrix}\begin{bmatrix} T_2(x, x) & R_2(x, x) \\ 0 & 0 \end{bmatrix}$$
$$+ \begin{bmatrix} 0 & R_2(x, x) \\ 0 & T_2(x, x) \end{bmatrix}\begin{bmatrix} 0 & 0 \\ \mathscr{R}_1(x, x) & T_1(x, x) \end{bmatrix}, \quad \text{(A-5)}$$

APPENDIX A 229

$$\mathbf{T}_4(x, x) = \begin{bmatrix} B'''(x) & A'''(x) \\ D'''(x) & C'''(x) \end{bmatrix} + 3 \begin{bmatrix} B''(x) & 0 \\ D''(x) & 0 \end{bmatrix} \begin{bmatrix} T_1(x, x) & R_1(x, x) \\ 0 & 0 \end{bmatrix}$$

$$+ 3 \begin{bmatrix} 0 & R_1(x, x) \\ 0 & T_1(x, x) \end{bmatrix} \begin{bmatrix} 0 & 0 \\ D''(x) & C''(x) \end{bmatrix} + 6 \begin{bmatrix} 0 & R_1(x, x) \\ 0 & T_1(x, x) \end{bmatrix} \begin{bmatrix} 0 & 0 \\ D'(x) & 0 \end{bmatrix}$$

$$\cdot \begin{bmatrix} T_1(x, x) & R_1(x, x) \\ 0 & 0 \end{bmatrix} + 3 \begin{bmatrix} B'(x) & 0 \\ D'(x) & 0 \end{bmatrix} \begin{bmatrix} T_2(x, x) & R_2(x, x) \\ 0 & 0 \end{bmatrix}$$

$$+ 3 \begin{bmatrix} 0 & R_2(x, x) \\ 0 & T_2(x, x) \end{bmatrix} \begin{bmatrix} 0 & 0 \\ D'(x) & C'(x) \end{bmatrix}$$

$$+ 3 \begin{bmatrix} 0 & R_1(x, x) \\ 0 & T_1(x, x) \end{bmatrix} \begin{bmatrix} 0 & 0 \\ \mathscr{R}_1(x, x) & 0 \end{bmatrix}$$

$$\cdot \begin{bmatrix} T_2(x, x) & R_2(x, x) \\ 0 & 0 \end{bmatrix} + 3 \begin{bmatrix} 0 & R_2(x, x) \\ 0 & T_2(x, x) \end{bmatrix}$$

$$\cdot \begin{bmatrix} 0 & 0 \\ \mathscr{R}_1(x, x) & 0 \end{bmatrix} \begin{bmatrix} T_1(x, x) & R_1(x, x) \\ 0 & 0 \end{bmatrix}$$

$$+ \begin{bmatrix} T_1(x, x) & 0 \\ \mathscr{R}_1(x, x) & 0 \end{bmatrix} \begin{bmatrix} T_3(x, x) & R_3(x, x) \\ 0 & 0 \end{bmatrix}$$

$$+ \begin{bmatrix} 0 & R_3(x, x) \\ 0 & T_3(x, x) \end{bmatrix} \begin{bmatrix} 0 & 0 \\ \mathscr{R}_1(x, x) & T_1(x, x) \end{bmatrix}. \quad \text{(A-6)}$$

The incremental coefficients for the source functions $g(x, x + \Delta x)$ and $h(x, x + \Delta x)$ are not difficult to find, and the fourth-order expansions are given below where

$$\begin{bmatrix} g(x, x + \Delta x) \\ h(x, x + \Delta x) \end{bmatrix} = \begin{bmatrix} g_0(x, x) \\ h_0(x, x) \end{bmatrix} + \begin{bmatrix} g_1(x, x) \\ h_1(x, x) \end{bmatrix} \Delta x + \begin{bmatrix} g_2(x, x) \\ h_2(x, x) \end{bmatrix} \frac{\Delta x^2}{2}$$

$$+ \begin{bmatrix} g_3(x, x) \\ h_3(x, x) \end{bmatrix} \frac{\Delta x^3}{6} + \begin{bmatrix} g_4(x, x) \\ h_4(x, x) \end{bmatrix} \frac{\Delta x^4}{24}. \quad \text{(A-7)}$$

COUPLED MODES IN PLASMAS, ELASTIC MEDIA, PARAMETRIC AMPLIFIERS

The subscripts have the same meaning as for the transmission matrix.

$$\begin{bmatrix} g_0(x, x) \\ h_0(x, x) \end{bmatrix} = \begin{bmatrix} 0 \\ 0 \end{bmatrix}, \tag{A-8}$$

$$\begin{bmatrix} g_1(x, x) \\ h_1(x, x) \end{bmatrix} = \begin{bmatrix} e(x) \\ f(x) \end{bmatrix}, \tag{A-9}$$

$$\begin{bmatrix} g_2(x, x) \\ h_2(x, x) \end{bmatrix} = \begin{bmatrix} e'(x) \\ f'(x) \end{bmatrix} + \begin{bmatrix} T_1(x, x) & R_1(x, x) \\ \mathscr{R}_1(x, x) & \mathscr{T}_1(x, x) \end{bmatrix} \begin{bmatrix} e(x) \\ f(x) \end{bmatrix}, \tag{A-10}$$

$$\begin{bmatrix} g_3(x, x) \\ h_3(x, x) \end{bmatrix} = \begin{bmatrix} e''(x) \\ f''(x) \end{bmatrix} + \begin{bmatrix} T_1(x, x) & R_1(x, x) \\ \mathscr{R}_1(x, x) & \mathscr{T}_1(x, x) \end{bmatrix} \begin{bmatrix} g_2(x, x) \\ h_2(x, x) \end{bmatrix}$$
$$+ \begin{bmatrix} B'(x) & A'(x) \\ D'(x) & C'(x) \end{bmatrix} \begin{bmatrix} g_1(x, x) \\ h_1(x, x) \end{bmatrix} + \begin{bmatrix} T_1(x, x) & R_1(x, x) \\ \mathscr{R}_1(x, x) & \mathscr{T}_1(x, x) \end{bmatrix}$$
$$\cdot \begin{bmatrix} 0 & A(x) \\ D(x) & 0 \end{bmatrix} \begin{bmatrix} g_1(x, x) \\ h_1(x, x) \end{bmatrix} + \begin{bmatrix} B'(x) & A(x) \\ D'(x) & C(x) \end{bmatrix} \begin{bmatrix} e(x) \\ f'(x) \end{bmatrix}, \tag{A-11}$$

$$\begin{bmatrix} g_4(x, x) \\ h_4(x, x) \end{bmatrix} = \begin{bmatrix} e'''(x) \\ f'''(x) \end{bmatrix} + 3\begin{bmatrix} B''(x) & 0 \\ D''(x) & 0 \end{bmatrix} \begin{bmatrix} g_1(x, x) \\ h_1(x, x) \end{bmatrix} + 3\begin{bmatrix} B'(x) & 0 \\ D'(x) & 0 \end{bmatrix} \begin{bmatrix} g_2(x, x) \\ h_2(x, x) \end{bmatrix}$$
$$+ 6\begin{bmatrix} 0 & 0 \\ D'(x) & 0 \end{bmatrix} \begin{bmatrix} g_1(x, x) \\ h_1(x, x) \end{bmatrix} + 3\begin{bmatrix} 0 & 0 \\ D(x) & 0 \end{bmatrix} \begin{bmatrix} g_2(x, x) \\ h_2(x, x) \end{bmatrix}$$
$$+ \begin{bmatrix} B(x) & 0 \\ D(x) & 0 \end{bmatrix} \begin{bmatrix} g_3(x, x) \\ h_3(x, x) \end{bmatrix} + 3\begin{bmatrix} 0 & R_3(x, x) \\ 0 & T_3(x, x) \end{bmatrix} \begin{bmatrix} 0 & 0 \\ D(x) & 0 \end{bmatrix}$$
$$\cdot \begin{bmatrix} g_1(x, x) \\ h_1(x, x) \end{bmatrix} + \begin{bmatrix} 0 & R_3(x, x) \\ 0 & T_3(x, x) \end{bmatrix} \begin{bmatrix} e(x) \\ f(x) \end{bmatrix} + 3\begin{bmatrix} 0 & R_2(x, x) \\ 0 & T_2(x, x) \end{bmatrix}$$
$$\cdot \begin{bmatrix} e'(x) \\ f'(x) \end{bmatrix} + 3\begin{bmatrix} 0 & R_1(x, x) \\ 0 & T_1(x, x) \end{bmatrix} \begin{bmatrix} e''(x) \\ f''(x) \end{bmatrix}. \tag{A-12}$$

The incremental coefficients can be calculated within the program by programming the preceding equations. The values of $A(x)$, $B(x)$, $C(x)$, and $D(x)$ and the first three derivatives must be defined within the program; the machine can then find the incremental coefficients.

APPENDIX B

Sample Computer Program

The recursive method of solving a coupled mode problem is perhaps best illustrated by a computer program. The following program was written for an SDS SIGMA 7 digital computer. The problem to be solved is the pendulum problem discussed in Chapter 8. No attempt has been made to write an efficient program. The program is written in the FORTRAN language.

FORTRAN PROGRAM

```
C     CALL FOR DOUBLE PRECISION
      IMPLICIT REAL*8 (A—H,O—Z)
C     DIMENSION THE ARRAYS
      DIMENSION T(2),R(2),SR(2),G(2),H(2)
C     SET THE STEP SIZE
      DEL1=0.01DO
      DEL2=DEL1*DEL1/2.DO
      DEL3=DEL2*DEL1/3.DO
      DEL4=DEL3*DEL1/4.DO
C     SET THE INITIAL CONDITIONS
      R(1)=0.DO
      T(1)=1.DO
      ST(1)=T(1)
      SR(1)=R(1)
      G(1)=R(1)
      H(1)=R(1)
      VO=0.DO
      UO=2.63DO
      UX=UO
      VX=VO
C     SET THE PRINT OUT INCREMENT—K*DEL1
      K=50
C     INCREMENTAL COEFFICIENTS
      ADLST=1.DO
      ADLT=1.DO+0.3DO*DEL1+0.09DO*DEL2+0.027DO*DEL3+0.0081DO*DEL4
      ADLR=DEL1+0.3DO*DEL2+0.09DO*DEL3+0.027DO*DEL4
      ADLSR=0.DO
```

```
      C   RECURSIVE EQUATION
          DO 1000 J=1,1001
          SX=DSIN(UX)
          CX=DCOS(UX)
          F=SX−0.5DO
          F1=VX*CX
          F2=−0.3DO*VX*CX−SX*CX+0.5DO*CX−VX*VX*SX
          F3=0.09DO*VX*CX+0.3DO*CX*SX−0.15DO*CX+0.3DO*VX*VX*CX
         1−VX*CX*CX−1.5DO*VX*SX+0.6DO*VX*VX*SX
          ADLH=F*DEL1+(0.3DO*F+F1)*DEL2+(0.09DO*F+0.6DO*F1*F2)*DEL3
         1+(0.027DO*F+0.027DO*F1+0.9DO*F2+F3)*DEL4
          ADLG=F*DEL2+(0.3DO*F+2.DO*F1)*DEL3+(0.09DO*F+0.9DO*F1+3.DO
         1*F2)*DEL4
          DEN=1.DO−R(1)*ADLSR
          R(2)=ADLR+ADLT*R(1)*ADLST/DEN
          T(2)=T(1)*ADLT/DEN
          ST(2)=ST(1)*ADLST/DEN
          SR(2)=SR(1)+T(1)*ADLSR*ST(1)/DEN
          G(2)=ADLG+ADLST*(G(1)+R(1)*ADLH)/DEN
          H(2)=H(1)+T(1)*(ADLH+ADLSR*G(1))/DEN
      C   CALCULATE UX AND VX
          VX=(VO−SR(2)*UO−H(2))/T(2)
          UX=ST(2)*UO+R(2)*VX+G(2)
      C   RESET THE COEFFICIENTS FOR THE NEXT STEP
          R(1)=R(2)
          SR(1)=SR(2)
          T(1)=T(2)
          ST(1)=ST(2)
          G(1)=G(2)
          H(1)=H(2)
      C   CHECK TO DETERMINE IF PRINTOUT IS DESIRED
          IF (J−K) 1001,1002,1002
     1001 GO TO 1000
     1002 WRITE (6,1003) J,UX,VX
     1003 FORMAT (I4,2D24.16)
      C   ADVANCE K BY 50 FOR THE NEXT PRINT
          K=K+50
     1000 CONTINUE
          END
```

This program will print out all values of $u(x)$ and $v(x)$ for $0 \leq x \leq 10$ at points 0.5 apart. It was used for the pendulum problem discussed in Chapter 8.

Author Index

Numbers in parentheses indicate the numbers of the references when these are cited in the text without the name of the author.

Numbers set in *italics* designate those page numbers on which the complete literature citations are given.

Adams, R., (4) 9, 23, 30, *44*, 146, 149, 216
Altman, J. L., 134, *150*

Bailey, P., 50, *76*
Bellman, R., 4, *9*, 23, *44*, 47, *76*, 98, *101*, 102
Bradley, Jr., W. T., 170, *181*
Brillouin, L., 39, *44*
Budden, K. G., 182, *200*

Carlin, H., 10, 15, *22*
Carr, H., 98, *101*
Close, C. M., *22*
Collin, R. E., 144, *150*
Cook, G., 98, *101*

Davies, H. A., *45*, 182, *200*
Davis, J. S., 227
Denman, E. D., (4) 9, 23, 30, *44*, 98, *101*, 146, 149, 216
DeRusso, P. M., *22*
Desoer, C., (4) *9*, *22*, 74, *76*
Dodd, R. E., 151, 162, *169*

Ewing, W. M., 162, *169*

Friedland, B., 5, *9*
Froberg, C. E., 106, *111*, (132), *133*

Gallawa, R. L., (205) *216*
Gerjouy, E., 201, *216*
Giordano, A., 10, 15, *22*

Hildebrand, F. B., (132), *133*
Howell, H. B., *76*

Jacobowitz, H., *76*
Jardetsky, W. S., 162, *169*

Kagiwada, H., 98, *101*
Kalaba, R., 4, *9*, 23, *44*, 47, *76*, 98, *101*, 102
Kaplan, L. J., 5, *9*
Kelley, J. P., 45
Kittel, C., 170, *181*
Kolsky, H., *169*
Kritz, A. E., *216*

Letton, III., W., 201, *216*
Li Pow So, 5, *9*
Lockett, J., 98, *101*
Louisell, W., 10, *22*
Love, A. E. H., *169*

Martz, C. W., 120, *122*
Melsa, J., *22*
Mintzer, D., *216*
Mizuta, H., 23, *44*

National Science Foundation, 8
N. B. S. Staff, *101*, 133

Officer, C. B., *169*
Oster, L., *216*

Pierce, J. R., 17, *22*
Poeverlein, H., 201, *216*
Porter, W., 5, *9*
Preisendorfer, R., 4, *9*
Press, F., 162, *169*

Ramo, S., 149, *150*
Ratcliffe, J. A., 182, *200*
Redheffer, R., 4, *9*, 23, *44*, 47, *76*
Reid, W. T., 23, *44*, 47, *76*
Roy, R. J., *22*

Sage, A., 5, *9*
Schmeidler, W., *76*

Schultz, D. G., *22*
Sethares, J. C., 170, *181*
Shanks, E. B., 91, *101*
Shimizu, A., 23, *44*
Smirnov, A. D., 6, *9*, 77, 83, *101*
Stock, D. J. R., 5, *9*
Struble, R., (4), *9*

Tai Ju Wei, 5, *9*
Twomey, S., *76*

Van Duzer, T., 149, *150*

Wang, P., 5, *9*
Wang Yu Yin, 5, *9*
Wing, G. M., 4, *9*, 23, *44*, 50, *76*, 114, *122*
Whinnery, J. R., 149, *150*
White, J. E., *169*

Zadeh, L., (4), *9*, 22, 74, *76*

Subject Index

ABCD parameters, 15
Adams-Moulton numerical, 7, 124
adjoint equations, 61
Airy functions, 6, 77, 83
Alfvén wave, 201
Appleton-Hartree formula, 183
artificial resonance, 30, 148

Bessel functions, 6, 83
Booker quartic, 191
boundary value problem, 57, 102

complex plane, 112
composite coefficients, 28
computer program, 231
constants, 154, 163, 175
    elastic, 163, 176
    exchange, 174
    Lame's, 154
    magnetoelastic, 175
control theory, 7
critical length, 100

down-conversion gain, 222

eigenvalue, 100
elastic waves, 151
electron collision frequency, 183
Epstein profile, 182, 185, 210
error function, 6, 87
exchange constant, 174

ferrite, 170, 174
forcing function, 41, 59, 91
Fortran, 231
Fresnel integral, 120
fundamental matrix, 39, 53, 73

gain, 222, 223, 227
gyromagnetic, ratio, 174

homogeneous differential equation, 47
hydrodynamic equation, 202

impedance, 24
impulse function, 81, 95
incremental coefficients, 6, 35, 40, 65, 173, 228
inhomogeneous differential equation, 60
initial value problem, 57, 89
integral, 6, 86
invariant imbedding, 7, 23, 152
inverse Laplace transform, 98
isotropic medium, 146
iterative, 104

Kirchhoff's mesh equation, 81, 94

Laguerre-Gauss quadrature, 98
Laplace transform, 20, 93
linear boundary-value problem, 57, 102

magnetization, 171
magnon wave, 170
matrix product, 39, 53
modulus, 154
    bulk, 154
    Young's, 154

negative resistance gain, 222
nonlinear boundary-value problem, 106
normal incidence, 146
normal mode, 10, 16

optimal control, 7

parametric amplifier, 217
partial pressure, 203
pendulum, 126
periodic solution, 72
phonon, 170
phonon loss constant, 175
plasma, 182
plasma equation, 202

Poisson's ratio, 154
polarization, 1, 191
potential, 157
  scalar, 157
  vector, 157
principle of invariance, 4

quartic, 191
quasilinearization, 7, 102

recursive equations, 23, 46, 51
refractive index, 148, 183
Riccati equation, 10, 46
Runge-Kutta numerical, 7, 91, 124

scattering matrix, 10, 134, 219
shear wave, 151
signal flow graph, 26
Snell's law, 160
spin relaxation, 175
state equation, 20

state transition matrix, 21
state variables, 20

Taylor series, 3
transient, 81
transmission matrix, 14, 25, 134, 141, 143, 219

unit step, 81, 94
up-conversion gain, 222

wave, 146, 154, 170, 185, 204
  electroacoustic, 204
  extraordinary, 185
  magnon, 170
  ordinary, 185
  phonon, 170
  pressure, 154
  shear, 154
  surface, 154
wave analogy, 1, 6

QA
372
D43

NOV 5 1971